Oliver Leistert, Isabell Schrickel (eds.)
Thinking the Problematic

Philosophy

Dedicated to all problems

Oliver Leistert is a media and technologies researcher at Leuphana University Lüneburg, Germany. He did his doctorate in media studies at University Paderborn and was a research fellow at Central European University in Budapest.
Isabell Schrickel is a doctoral researcher at Leuphana University. Her main areas of research are the history of science, environmental humanities and media studies.

Oliver Leistert, Isabell Schrickel (eds.)
Thinking the Problematic
Genealogies and Explorations between Philosophy and the Sciences

[transcript]

Bibliographic information published by the Deutsche Nationalbibliothek
The Deutsche Nationalbibliothek lists this publication in the Deutsche Nationalbibliografie; detailed bibliographic data are available in the Internet at http://dnb.d-nb.de

This work is licensed under the Creative Commons Attribution-NonCommercial-NoDerivatives 4.0 (BY-NC-ND) which means that the text may be used for non-commercial purposes, provided credit is given to the author. For details go to
http://creativecommons.org/licenses/by-nc-nd/4.0/
To create an adaptation, translation, or derivative of the original work and for commercial use, further permission is required and can be obtained by contacting rights@transcript-publishing.com
Creative Commons license terms for re-use do not apply to any content (such as graphs, figures, photos, excerpts, etc.) not original to the Open Access publication and further permission may be required from the rights holder. The obligation to research and clear permission lies solely with the party re-using the material.

© 2020 transcript Verlag, Bielefeld

Cover layout: Kordula Röckenhaus, Bielefeld
Proofread by Selena Class
Typeset by Justine Buri, Bielefeld
Printed by Majuskel Medienproduktion GmbH, Wetzlar
Print-ISBN 978-3-8376-4640-5
PDF-ISBN 978-3-8394-4640-9
https://doi.org/10.14361/9783839446409

Printed on permanent acid-free text paper.

Contents

Acknowledgements 7

**Introduction to Thinking the Problematic:
Decentring as Method and Ethos**
Oliver Leistert & Isabell Schrickel 9

The *Problems of Modern Societies* — Epistemic Design around 1970
Isabell Schrickel 35

The Problematic of Transdisciplinary Sustainability Sciences
Esther Meyer 69

**A Genealogical Perspective on the Problematic:
From Jacques Martin to Louis Althusser**
Jean-Baptiste Vuillerod 93

**'The problem itself persists': Problems as *Missing Links* between
Concepts and Theories in Canguilhem's Historical Epistemology**
Thomas Ebke 109

**Compositional Methodology:
On the Individuation of a Problematic of the Contemporary**
Celia Lury 127

**From Critique to Problems and the Politics of the In-act with Bergson,
Deleuze and James**
Christoph Brunner 153

Pragmatics of a World To-Be-Made
Martin Savransky..179

About the Authors...195

Acknowledgements

This book emerged from a workshop initiated by Erich Hörl, Oliver Leistert and Martin Savransky called *Thinking the Problematic*, held at Leuphana University Lüneburg in June 2017.[1] The workshop was as part of a research theme of the research project *CCP | Complexity or Control. Paradigms for Sustainable Development*, funded by the Volkswagen Foundation from 2015 to 2019. Most of the contributions to this book had been presented initially at the workshop. The editors thank all participants for shaping an exciting discussion: Didier Debaise, Thomas Ebke, Craig Lundy, Celia Lury, Patrice Maniglier, Esther Meyer, Dimitris Papadopoulos – and in particular Erich Hörl for sparking the discussion and Martin Savransky, who we also had the pleasure to host as a visiting fellow at our project. We also thank Jeremias Herberg, Gregor Schmieg and the whole CCP team for their curiosity, openness and patience for the topic, Ina Dubberke for the smooth institutional support, and Selena Class for rapid, but nonetheless thorough, proofreadings.

---

1 For a full line-up and the programme see https://www.leuphana.de/zentren/cgsc/aktuell/termine/ansicht/datum/2017/06/22/workshop-thinking-the-problematic.html.

Introduction to Thinking the Problematic: Decentring as Method and Ethos

Oliver Leistert & Isabell Schrickel

Our occupation with problems (and of problems with us), the entanglement of problems and ideas, and their relations with thought, concepts and solutions, the universality, generosity and violence of problems, and the continued problems we cultivate in order *not* to develop a sense of problems, a sense that would affirm their transformative offerings and expose us to a risk – these are the topics that the contributions to this volume revolve around in rather different spins. The book is a contribution to the problem of how, when, where and why problems matter, and for whom, and therefore to the inescapable and unmistakable catastrophic resonances that are occuring when modern societies continue to cultivate their *amor fati* with false problems, 'that are only possible through various confusions between terms that had been previously separated and constructed, but whose modes of construction are no longer put into question', as Didier Debaise recalls (2018: 20). *Thinking the problematic* might therefore as well mean an endeavour for decentring our thinking in order to think again, and to put the modes of construction of problems into question. This sounds quaint, as common sense has it that thinking is obviously part of everyday life. But when we look beyond the cognitive activity as such and understand thinking as a process in and after which a difference has been made – and this difference does not entail, for the moment, any limitations – it turns out that neither common sense nor everyday life help us to engage in the process of thinking. Quite to the contrary: their role is to stabilise, to make certain and to establish continuity – a sound milieu for false problems to flourish in.

The force of thinking to transform what it has captured is thus the topic here, and as such it is one way to explore what a problematic might turn out to be – a *positive* conception of a problem. Most of us know this force from events that shook us and had an impact on how we situate ourselves in the world. In retrospect, however, the actual problematic tends to hide within historical narratives of progress that value the solutions of problems, but not their original stating. Many branches of science and discourses of science and technology, especially in their instrumental, solutionist and result-oriented reasonings, are still subject to this constraint.

The term problematic is not fixed, and has never been.[1] There are, in fact, significant variations in its use and description that prove the vitality of the term or, bluntly stated, its existence as a force on the plane of immanence, as Gilles Deleuze might have it. Whereas some philosophers, scientists, activists and thinkers refer to problems, and tend to address a problematic, others refer to a problematisation and focus on an activity – the construction of a problem. In addition, an important strand of problematisation refers to ontology and ethos, to the living and how to live. Indeed, turning to the problematic implicates us in the problematisation of ontologies of thought/thinking,[2] a paradoxical phrase at first glance, as Western cultures tend to separate thinking and being, leading to a dramatic devaluation of ontology as a field of thought in general. The division of the two has enshrined ontology as being primarily studied in academic ivory towers by experts, without further consequences than a thesis without a readership.

In light of this domestication of problems we attempt to contribute to a more recent intellectual engagement with several original and critical contributions to a positive understanding of problems and the problematic, cultivated primarily in the 20th century French philosophical and epistemological traditions. In contrast to the various negative concepts of problems that are prevalent in particular disciplines or other philosophical traditions – problems as cognitive obstacles, as a relation between the known and the

---

1 For an etymological definition of the word *problem*, see Schrickel, this volume, p.50.
2 The historicisation of ontology gained profound traction in a truly pluralistic perspective not long ago when anthropologists started to study ontologies in comparative ways without recasting alien concepts onto abstract modern terms. Although the beginning of these efforts can be dated back to the 1980s, considering for example Marilyn Strathern's *The Gender of the Gift* (1988), it is only recently that it was expressed programmatically with Charbonnier et al.'s *Comparative Metaphysics: Ontology After Anthropology* (2016). See also Viveiros de Castro's *Cannibal Metaphysics* (2014) for a sense of the intricacies of a truly pluralist universe and the role of concepts therein. For a history of the concept of problems in the history of philosophy from antiquity onwards, see Bianco (2018).

unknown, or as a conflict between different ideas for instance (Maniglier 2019) – the authors of this volume engage with philosophers, activists and historical contexts of the problematic that questioned the prevailing passive, ahistorical, deficient and solution-oriented character of the notion of the problem in many ways, called for a break-up of the problem-solution coupling and argued for problematisation as a process of transformative engagement. Taking a particular intellectual ethos in the French philosophical and epistemological tradition, where problems have been understood as a truly creative and intrinsically productive force, as a starting point, this volume attempts to trace the problematic throughout a variety of authors and cases, through philosophy, epistemology and a series of practical endeavours. We seek to trace both the genealogy of thinking the problematic and the seeds of these intellectual projects in discourses around inter- and transdisciplinarity, the scientific orientation towards 'real-world problems' and the 'problems of modern societies', and the role of the concept in the histories of systems thinking, public planning and sustainability science. Especially at times when science policy is so heavily geared towards big problems and grand challenges – public health, global sustainability or the adoption of artificial intelligence – it seems apt to problematise, historicise and complicate the problematic anew.

With this project we built on the previous achievements of a number of workshops, discussions and publications that picked up the threads of the problematic in recent years. The research project *Transdisciplinarity and the humanities: Problems, methods, histories, concepts* (2011-2013) at Kingston University London noticed – quite similar to our experience at CCP – also the lack of theoretical work on the concept of the problem and dedicated their first workshop, *From Science and Technology Studies to the Humanities* (2012), to the concept. Peter Osborne observed that although transdisciplinarity as a research methodology is broadly oriented towards the collaborative solution of societal problems, such as environmental sustainability and health and problems in the 'life-world' (Hirsch Hadorn et al. 2008), it seems entirely unclear what a problem is. Is it 'something that requires the positing of practical solutions, or is a problem, primarily, something that defines a shared field of inquiry (a problematic), the investigation of which may take radically unexpected turns, leading to a reproblematisation – critical or otherwise – of the original issue?' (Osborne 2015: 13). Since the programmatic of a practical rationality of states or state-like entities as organisers and sponsors of this kind of research will certainly want to maintain control over the form of the process of disbursement, and ensure accountability and applicability, there is a systemic preference for solutions to the detriment of the process of problematising what is actually at stake. Thus, he concludes, inter- and transdisciplinarity have lost the more radical socio-political content associated with the rise of these movements in the years around 1970. Osborne and his colleagues then propose to involve European 'theory' (French theory, German critical theory, literary criticism) in transdisciplinary research, as they provide approaches to reflexively iterative processes of problem definition, investigation and reformulation.[3] The problematic was also recently the subject of a special issue of *Angelaki*, edited by Sean Bowden and Mark G.E. Kelly, summoning some of the finest minds to produce new connections or differences among the canonical and the less canonical French epistemologists and philosophers that have enriched the discourses in the humanities and other disciplines in the 20th century in unprecedented ways.[4] Martin Savransky also edited an exciting collection of papers for a special issue of *Theory, Culture & Society* on the problematic, with which many of our interests resonate, and some of which we will return to later in this introduction.

This volume attempts to open up the problematic, too. The contributions of Esther Meyer and Isabell Schrickel, in particular, trace the critical productivity of the concept in different historical, scientific and practical contexts and add to the problematics of inter- and transdisciplinarity. Jean-Baptiste Vuillerod and Thomas Ebke return to the genealogies and structural functions of this term in French theory. Celia Lury composes a methodology for the individuation of a problematic of the contemporary. Christoph Brunner and Martin Savransky suggest operative building blocks for the cultivation of situations that harness the transformative powers of problems. To engage with different problematics here then addresses the limits of our thinking, too, by offering different accounts from a variety of fields that, surprisingly enough, to date have never been assembled in one book. We have found ourselves in dialogue more than once during the finalisation of this work-

---

3   In the same winter of 2012, another workshop at Goldsmiths College in London critically mobilised in a similar manner the works of Gilles Deleuze and Félix Guattari in particular to discuss the problem of transdisciplinarity and the problematic dynamics to re-disciplinarise and re-establish itself on a transcendent element (see Collett 2019).

4   For a brief overview of all contributions see Bowden/Kelly 2018.

shop's outcomes regarding the impossibility of determining the limits of the problematic, and take this as an encouraging detail of its relevance in a genealogical perspective. It turned out, after we provisionally ended our conversations, that it remains an open project to thoroughly look beyond the more recent receptions and interest that the problem of the problematic has received.

Lineages of problems and problematisation

The history and philosophy of science is rich with famous problems being solved and has provided a great variety of strategies of problem-solving: abstraction, analogy, divide and conquer, hypothesis testing, lateral thinking, proofs, trial and error, or workarounds – numerous tools and approaches to overcome problems have been developed throughout history. Problems solved assure us in often anecdotal ways of the constant progress in modern science, and problems unsolved are seen as epistemic puzzles that are being confronted with confidence and faith in future problem-solving capacities. In a positivist concept of science as a properly demarcated and ahistorical endeavour problems function as some kind of placeholder for the time span needed to find the solution. Problems are obstacles to be removed, means to test specific solutions, they are negative states of uncertainty, ignorance and methodological imperfection bound to dissipate with the solutions that scientific and technological progress yield. Consequently, traditions like logical positivism rejected the 'great questions': philosophical, metaphysical, vital and singular problems are in fact *Scheinprobleme* (Carnap 2005 [1928]) – pseudoproblems – which are incapable of solution not because of their profundity but because they pose nothing to be solved.

On the one hand we could simply acknowledge the fact that these traditions drew the limits of scientific jurisdiction and the boundaries of scientific and non-scientific disciplines – in their case between physics and philosophical metaphysics, Freudian psychoanalysis or Marxist social criticism – so neatly and sorted out their scope and area of responsibility in quite transparent – yet polemical – ways. But also, the solutions derived from such neatly demarcated scientific fields will always reach beyond. Solutions come into existence as theoretical perspectives, as socio-technical arrangements and pathways, as products and services. Solutions become effective by bringing together concepts, objects, tools, techniques, scientists, institutions and publics in new ways. Sometimes, solutions consolidate the problem by deepening the goals and values already visible as the basis on which the problem emerged, and sometimes solutions open up paths for transformations and alternative futures. There is always some excess in solutions, as they could have been otherwise. Thus, solutions are always more than scientific – as they are always already problematic, too. For a long time, the history and philosophy of science did not pay much attention to either the notion of the problem or the solution. One will search in vain for comprehensive entries on these lemmata in encyclopaedias of philosophy or science, and their reach beyond colloquial meanings and explorations of these operational terms even today (Mittelstraß et al. 2005-2016; Serres/Farouki 1997; Lecourt 2006). This is astonishing, not least as we have come to acknowledge for a long time now that we are indeed surrounded and impregnated by scientific applications and products, embedded in infrastructures and policy cultures that are based on scientific expertise and technological solutions that our societies co-evolve with.

It has been widely recognised that the French epistemological tradition, which established itself over several generations in close examination and discussion with contemporary science, has provided essential perspectives and new avenues to engage with modern science and its problems and the role of knowledge in society more broadly. The struggles over epistemology in France during the 1960s, for example, are evaluated today as instances of important mutual exchanges between the sciences, philosophy and society, providing novel techniques and tools for argumentation, thought and action, and a specific mode to reflect on the role of science in society (Erdur 2018). These epistemological, philosophical and theoretical engagements became important undercurrents and intellectual resources in debates over inter- and transdisciplinarity that emerged during the late 1960s and that led to the establishment of new institutions, academic fields and approaches to solving real-world problems (Klein 2014). The subsequent rise of the various fields of historical, philosophical and social analysis of science during the 1960s and 1970s – science studies and the history and sociology of science and science and technology studies a little later – also had a close connection to, and drew major impulses for analysing and questioning processes of knowledge production and their role in public affairs from these engagements, which has been acknowledged until recently (Biagioli 1999; Biagioli 2001). And finally, the vast potential of these writings for a constructive critique of science policy and the prevalent organisation of problem-oriented transdisciplinary science has recently been rediscovered, as we have seen (Osborne 2015; Collett 2019; Maniglier 2019).

These strands are picked up by Meyer and Schrickel in their contributions to this volume. ESTHER MEYER provides a critical assessment of discourses and constructions of problems of sustainable development in recent transdisciplinary (td) sustainability sciences, and asks 'How can we think of methodologies for td sustainability research that are coherent with epistemologies of the problematic?' She suggests mobilising the philosophy of Gilbert Simondon, as he offers a 'radically transdisciplinary' alternative to the mechanical concept of *development* covered in the hegemonic versions of sustainable development, in particular through his theory of individuation, where a problematic arises as a resonance between an exteriority and an interiority. Meyer refers to several approaches in recent td sustainability research that take such an initial situation as a methodological starting point, including her colleagues and Meyer's own method of 'thinking practice of problematic designing'.

ISABELL SCHRICKEL offers in her contribution a historical account of an epistemic shift characterising the years around 1970, and discusses the symptomatic conjuncture of the notion of the problem in it. The rise of 'problem-talk' – from 'wicked problems' to the 'world problematique' – signifies a shift in epistemic sensibilities at the time, Schrickel argues, where new institutions and forms of knowledge were constructed around problems that would allow societies to change, to adapt, or to intervene in their futures. She does not suggest that there is a particularly strong connection between the writings of the authors subsumed under the label of French theory, with their nuanced approaches to the problematic, and, for instance, the simultaneous considerations of planning experts, systems analysts and bureaucrats from agencies such as the OECD, the Club of Rome and other institutions who put the 'problems of modern societies' on their agenda. Schrickel observes, however, that they share the idea of a positive conception of problems as intrinsically productive and transformative instances, and a sensibility for the lurking danger of instrumentalising problems, for example in order to maintain a status quo or to make particular policy options more likely than others. She embeds her observations in a broader historical analysis of the political situation and the academic landscape of those years, and dis-

historical epistemology as a watchguard of normativity in the scientific process, a political project, as Ebke concludes.

Towards an ethos

It is the very late Foucault who, in an interview with Paul Rabinow, takes up the concept of a problematic and, rather surprisingly, relates his works in the history of thought to the rediscovery of 'a general form of problematisations' (Foucault 1984: 389).[8] For him – and this is where Foucault provides a glimpse into the reconstruction of an ethos as opposed to a morality based on transcendental laws – problematisations are discernible within discursive responses to difficulties that are transformative in the sense that they react to and effect practical solutions. Problematisations are instigated by some uncertainty in a specific field and provoke simultaneously different, at times even contradicting, solutions. This explains why stating a problem is much more difficult than stating its solution, as Bergson put it in the context of speculative problems: a problematisation articulates difficulties in manifold ways and thereby develops the conditions under which possible responses can be given. This is a situated practice of thought, rich in context and seldom possible to reconstruct backwards, since the specific work of thought in the form of problematisations cannot be grasped after the fact, as a succession of representations, because 'while carrying out the work of thought under the experimental form of a historico-practical test imposed by our present', it is 'inseparable from the modes of problematisation our present makes us capable of' as Stengers (2019: 11) explains the immanent distribution of forces that at the same time impose and capacitate, or even capacitate by imposition. For Foucault, the ability to problematise turns out to be a condition of freedom, through which he probes a thoroughly positive problematic conception and a freedom freed from transcendental burden and authority.[9] Paul Macherey further suggests that Foucault's notion of thought is intrinsically connected to a manifestation of a limit, or an un-

---

8 Although it should be stated that he remains rather cautious by setting the phrase conditionally, as if he wanted to signal the impossibility of this endeavour.
9 How problematisations concern ethics and freedom in Deleuze's and Foucault's works has been analysed by Erinn Gilson (2014).

certainty, as 'the subject opens up for itself a domain of intervention, inside not outside the system, by taking the position from which a certain claim to freedom becomes meaningful' (Macherey 1998: 101). Here again, the positionality returns as a condition to thought, and the singular turns out to be of the universal ('in the system') as a condition for a transformation, whereas if it was of the general, freedom would, once again, become abstracted and thus float outside the system.

The Belgium philosopher, historian of science, activist and former chemist Isabelle Stengers has contributed to an actualisation of the problem-ethos nexus in two distinct manners: firstly, for a while now, together with Didier Debaise, Martin Savransky and others, she demonstrates how to apply pragmatistic concepts from the philosophy of William James as tools that can operate as instigators for problematic practices (see below, and Savransky and Brunner in this volume). And secondly, she recently took up Foucault's notion of problematisation as a form of 'transformative engagement'. As modification of 'the relation we entertain to our own reasons' (Stengers 2019: 3), she seeks an experimentation with consequences. Here, the method of application must emerge in the encounter with the problem, and the value of knowledge refers to one's own limits (see also Lury in this volume). This problematic shares the Deleuzian dramatisation of an idea to be actualised as a problem once it takes possession of its bearer, who is violently forced to think and becomes herself part of a thought as much as this becoming transforms the parts involved. The outcome, in the form of a new structure, is a hypothetical problem with its field of possible solutions, 'issued from the problematising power of the idea which selects and mobilizes what the problem needs in order to determine itself and to receive the solution it deserves' (2019: 7). Stengers proposes that the Deleuzian notion of an idea that has powers to insist and demand actualisation, but never fully exhausts actualisation, is what demarcates the problematising subject that is referenced by Foucault and whose truth is a demand by a transformation originating in practices. 'If modes of problematisation are formed on the basis of practices, they also relay the concerns whose insistence these practices manifested' (2019: 10), she writes, and the concept of relaying is one of those prolific enrichments by Stengers to the modes of the problematic. By nature practices are situated and by nature they are a diagnosis of their milieu, of what is possible, a test of concerns without judgement. Here, problems serve as tools for an ethopoiesis – the fabrication of a situationally limited ethics.

In addition, Stengers introduces Étienne Souriau (2015 [1943]) to the lineage of historians of problems, because his concept of 'questioning situations' that prey upon those who admit to them establishes an ontological risk in the form of a problematic, as the answer to the problem may remain insufficient, and simultaneously imposes a responsibility, as the problematisation must resist already existing, ready-made solutions (on Souriau, see also Savransky in this volume). Transformations instigated by such a risky situation may fail, which very much resembles James' concept of a genuine option (see Stengers 2009), while at the same time Souriau shares Deleuze's concept of the Idea as the bearer of the thinker, although in a 'less violent' tune, as Stengers explains (2019: 8).

From situated knowledges to the cultivation of situations

This ontological or epistemological positionality characteristic of the problematic is echoed many times in recent observations and proposals. Maybe (now) most prominently, and not that long after Foucault's death in 1984, Donna Haraway invested her thought (and anger) into the outline of a situated knowledge (1988) that in many ways, knowingly or not, resembles elements that are familiar from the works attributed by During and others to the historians of problems: embodied objectivity, limited location, partial perspectives and situated knowledge are proposals that ultimately concern an ethical practice in the form of an accountability based on webs of connections and the simultaneous interrelatedness of the epistemological, ontological, ethical and political planes. Reading Haraway's proposal today remains instructive because (amongst other reasons) one of its most prominent polemical antagonists is the spectre of relativism. Relativism figured as a discursive tool to devalue all self-limiting epistemologies as it sets up the false, but exclusive, binary between relativism and objectivism. Haraway rightly points out that 'the "equality" of positioning is a denial of responsibility and critical inquiry. Relativism is the perfect mirror twin of totalisation in the ideologies of objectivity. […] But it is precisely in the politics and epistemology of partial perspectives that the possibility of sustained, rational, objective enquiry rests' (Haraway 1988: 584).

Sadly, these polemics that position an unfettered objectivism on the one side and an unconstrained relativism on the other, continue to resonate up until today within discourses on the normative frameworks and scopes of the sciences. Setting up relativism as the other of objectivity is a perfect example of a false problem that only a scientist's reason could come up with in order to retain *his* exclusive and exhaustive access to truth. Today, these polemical attacks on what back then was called postmodernism are instances of powerful strategies to delegitimise any kind of problematisation that questions and limits scientific practices and knowledge productions. Positionality, in this polemic, equals relativism, an absurd rhetoric motivated by an authoritarian judgement struggling for legitimation. As it is evident that the disputes Haraway refers to are truly false problems, their many returns signify the political stakes inherent to them. At the core, it pertains to weakening the view that science is practice and facts are made, a product, and not an expression of nature herself, as the term 'laws of nature' still proposes. The purification and rhetorics of science as nature's language still has outspoken purchase in the battle for funding and self-legitimation. This continued immunisation strategy of scientific reason has been nurtured by, and entered into a new process of naturalisation with, the advent of today's data science, so called big data, algorithmic processing and what still, or again, is referred to as artificial intelligence. Here, the phantasma of a general, unsituated objectivity has re-emerged as digital data now get treated as splatters of the real.[10]

---

10 This recent and ongoing regression in scientific practices instigated by the abundance of data and cheap processing power increasingly reduces many branches of science to mere engineering tasks. While this development is not new per se, and, of course, Haraway was among the first feminists to address the capital-driven technologist attitude of science (2004 [1985]), what can now be observed all along formerly methodologically diverse fields is a reduction of diversity in science through the application of the same, often patented and thus black-boxed, bundles of algorithms, and partially even the same training datasets. Louise Amoore reports that 'scientific data begin to incorporate the emotional, affective, and speculative domains, while, on the other hand, knowledges considered to be "non-scientific" are authorized as science. [...] the degrees of doubt always already present within mathematical probability multiply and take flight as imaginable, if not strictly calculable, possibilities' (2013: 10). Such a 'speculative' calculus attempts to objectify – or reify – the virtual by replacing it with the possible a computer can process. This operation of capture extends the reach of formalised methods beyond probabilities, the episteme of modern societies, into the realm of possibilities whose only limit is computability itself, therefore constructing an unlimited upgradable plane. The prospects for a feminist data science (D'Ignacio/Klein 2020), for instance, however reasonable in itself and well intended, carry the burden of possibly turning out to function more as a vindication than a cure.

Haraway, intervening into this polemical debate against postmodernism, unambiguously drew the line for any claims to objectivity in the necessity of partiality, because in return this retains and cultivates plurality and diversity. This obliging relation continues to form, up until today, the conditions of the possibilities of knowledges that a subject can relate to herself, even when the grounds appear to have shifted today: 'Positioning is, therefore, the practice in grounding knowledge [...] Positioning implies responsibility for our enabling practices' (1988: 587). Haraway later (2008: 71) rephrased this ethical backstop as 'response-ability', which bears a more positive conception that at the same time is scaled down to a subject's dimension of apprehension: a pragmatic care of the problematic.

Situating objectivity with partial perspective, and with what is of importance, resonates well with Didier Debaise's problematisation of 'the bifurcation of Nature'. He is showing, with recourse to Alfred North Whitehead, that scientific reasoning has taken the position of nature's original expressions, masking thereby in a second operation the rich pluralism inherent in nature, as nature is reduced to the limitations of a scientific axiomatic and localisable matter within an absurd reductionist concept of time. This leads to severe confusions 'where everything is reversed: operations replace ontology, and abstraction replaces the concreteness of things, and the possibility of the knowledge of existence in itself' (Debaise 2017: 26). To 'take the tool for the universe' lets thought oscillate freely in false problems, between 'primary' and 'secondary' qualities, of which 'all of the divisions between beings, all the oppositions between their attributes and their aspects, are derived: existence and value; real nature and apparent nature; fact and interpretation' (Debaise 2017: 2). What is more, the reification of this bifurcation effectuates a delegitimation of other metaphysics. Only scientific reason has access to the real, causing 'a desertification of all modes of existence: the reduction of mental beings to simple representations, of fictions to imaginary realities, of values to subjective projections onto nature' (Debaise 2018: 22). Maybe the late Foucault sensed this power of desertification when he felt the obligation to the archaeological and genealogical restitution of practices of care from antiquity in his history of sexuality after the first volume.

On the real problem of data justice – in contrast to the false problem of data ethics – see, in an explorative manner, Dencik et al. 2019.

In any case, against these 'active anesthesia of thought' (Debaise 2018: 23) that domesticated the problematic as a problem-solution calculus of scientific reason, a fresh take on the restitution of the relevance of experience in a minor tune continues to spread. By way of setting up obligations in the form of pragmatism's 'genuine options' (William James), any claims by abstractions to an exclusive access to truth are undercut and rendered impossible. This way, the concept of truth undergoes a massive reform, as truth now signifies the ability to convey from within a situation all the constraints necessary. This way, truth and present converge – whereas scientific reason would separate from without (or from God's perspective, as Haraway called it) all that is inside and therefore unfit for claims on truth. Truth becomes inclusive as it excludes any reach beyond its situational present. Programmatically, it 'enrages any majority thinking' (Stengers/Debaise 2017: 19), as it subverts and annihilates the authoritarian grip on the distribution of truth. This pragmatist reformulation of truth has been embedded within many exercises and narratives for the cultivation of problems.[11]

In this vein, MARTIN SAVRANSKY, in his contribution, returns to James' concept of a 'fringe' that constitutes a vector of indetermination in thought, acting as a generative force of the problematic. Speculating on the title of our book, *Thinking the Problematic*, Savransky points to the paradoxes contained therein, as he suggests that in it thought folds back on itself. With reference to Deleuze's deconstruction of the representational image of thought, Savransky narrates how problems have withered into an epistemic obstacle to be overcome under the reign of instrumental reason – a matter of puzzle-solving, amounting to an impossible attempt to exhaust the problematic with one universally valid reason. For Savransky, thinking the problematic means to learn how to sustain and entertain the insistent possibility contained within a problematic. For this, he returns to Souriau's ontology of intensities, because it problematises heterogenesis. Souriau exemplifies heterogenesis with sculpting, as the statue is a generative problem that turns the sculptor into its means. Intensification thus involves metamorphosis of a work done, Savransky argues, and this leads him to speculate how to conjure the problematic, and to look for arts and practices of other modalities of truth speaking, such as the oracle's practice of veridiction that demands

---

11 Such exercises can be found, for instance, in *Breaking the Spell* (Pignarre/Stengers 2011) or related works (Savransky 2016; Stengers 2015).

a transformative response by the consultee. Thinking the problematic, he concludes, may, rather simply, amount to trusting the possible for its generativity.

For Savransky, Stengers, and likewise for many other authors mentioned in this introduction – in many ways also for Michel Foucault – a productive source of reasoning about problems remains one specific exercise that sets out to perform the transformative arts of the problematic without restraint. The anti-representational thought brought about by Gilles Deleuze's *Difference and Repetition* (1994 [1968]) stands out in rigour and generosity (as does *The Logic of Sense* (1990 [1969]). The continuity of Deleuze's formative works within the more recent literature on the problematic prevails, because Deleuze most explicitly formulates a genuinely positive concept of problems, which situates them 'on the side of events, affections, or accidents rather than on that of theorematic essences' (187). Further, Deleuze, in a truly original style, has deconstructed and unmade the bifurcation of nature as he shows the conceptual poverty it produced. A careful reception of these works of Deleuze taps into a richness in problematic thought that remains unmatched, especially when considering chapters 3 and 4 of *Difference and Repetition*, where Deleuze presents the problem as a qualifier of ontological relevance. 'The problem of thought is tied not to essences but to the evaluation of what is important and what is not' (189). This echoes the Whiteheadean metaphysical ethos of 'asserting importance as a primary category of the experience of nature' (Debaise 2018: 25). If the problematic maintains importance, meaning both being important and opening the senses to what is important, then it retains a generativity or inventiveness that takes hold of bodies and minds alike. This possibly violent force is full of surprises and difficult, if not impossible, to govern without losing its grip – that is, its importance. Problems are in correspondence with, to and from, norms and normativity, as they instigate new practices that test and individuate the milieu they are positioned in. Their primary operation to decentre and change not only targets perspectives and positions but axiomatics, too – these order-codings of constructed necessity delimiting all that is possible. While this may sound pathetic, it should be stressed that the activity of problematisation is discursive *and* subjective, molar *and* micro: we can find axiomatisations all around, whose function it is to continuously discriminate between ground and figure, to enable scales that themselves enable units of measure, and this way provide the necessary means for the implementation of norms and normativity. Put differently, by way of problematisations we actively un-categorise the categorised and tap into the 'chaosmosis', as Félix Guattari (1995) has named this generic mess in his unprecedented conceptual generosity. When some of our senses are positioned to dispose of false certainties generated by exclusive access to truth by scientific reasoning, our aptitude towards a pluralist reasoning and non-judgemental but inclusive concept of truth gains traction.

On such a plane *Christoph Brunner*, in this book, investigates the conditions for a politics constituted by the 'collactive', a collective relaying acts. He takes inspiration from the rejection of classical modes of critique by Stefano Harney and Fred Moton, who call for a new mode of critique that escapes the illusion of an autonomous oppositional subject and that refuses the common sense orderings of truth this subject is aligned with. Instead, it is in the movement of flight, in a durational concern, that the act lingers. In a confluence of a range of concepts from Bergson, Deleuze and James, Brunner distils a shared critique of common sense, before he turns to Bergson's method of intuition and Deleuze's take on it in order to turn it into a speculative-pragmatic process of problematisation aiming – through affirmation – at an invention of the present to overcome the present, a process of becoming relationally. An example he gives for a reconceptualisation of time is the Afrofuturist multiplication of temporalities. Problems as transversal operators effectuate in Brunner's praise for movement the possibility of an in-act, a slipping into the event without beginning but 'with a joy of entering the interplay of durations'. Ultimately, this resistance against the present turns to 'the inventive powers of shape-shifting that present intuitively'.

Problems are figured to belong to instigators of change and transformation, to pertain to the necessity to develop, at length and with precision also, in the works of Gilbert Simondon. The works he cites in his thesis that he defended in 1958 range from cybernetics to the pre-Socratic apeiron. Brian Massumi, himself a philosopher of problems, assumes that Simondon's ecological philosophy was intellectually inaccessible in most times, not only because it is only now being translated, but because it lacked a climate of openness towards ontological concerns in the 80s and 90s, when the long paradigmatic idea of constructivism 'was in fact unequal to the question of ontogenesis that it was called upon to take up by virtue of the juncture at which it found itself' (Massumi 2009: 37). The constructivists' own legitimation rested in an ontological disdain that can be considered as a discursive necessity of that time in order to theoretically posit social or cultural perspectives on things and their subject positions. 'Ontology, several generations of theorists were taught, was the enemy. Epistemology, which always carries ontological presuppositions of one kind or another, was at best a false friend. Finding a path to ontogenesis by unabashedly bringing the two together again, albeit in a new way, was simply inconceivable' (ibid). But there is more to Simondon's untimeliness. As his theory expresses complex becomings with only very few genetic concepts and without a general principle, he developed an 'integral inventivism' (Massumi) that equally concerns matter as it concerns thought and ideas – an impossible architecture of theory for constructivism and most of the humanities until recently.

This theory of qualitative change cuts radically through the world's distribution into disciplines – not only because a world divided into disciplines causes unsolvable epistemological obstacles for such a genetic endeavour, but, even more relevant, their founding principle to discriminate and order the real in their logic, this very abstraction, is causing the construction of disciplines that implicitly import normative assumptions. Simondon's sensibility here echoes his close knowledge of the works of his teacher Georges Canguilhem, who analysed the recurrent installation of the junction between the normal and the pathological in the sciences of the living. This spurred Simondon to reject psychology: 'The constitution of two spaces [the normal and the pathological] only expresses the essential bi-polarity of normativity for a psychological classification, and obfuscates the implicit sociology and social technics' (Simondon 2005: 270). Consequently, he refers to psychosociology in his theory to underline the necessary and inseparable relation of the interior and of the exterior for an individuation of beings.

A problem for Simondon is what 'resolves an anterior incompatibility through the apparition of a new systematic; what was tension and incompatibility becomes functional structure' (Simondon, quoted in Voss 2018: 100). This new functional structure, otherwise said, is the outcome of a formative process, initiated by a problem: 'To solve a problem is to be able to step over it, to be capable of recasting the forms that are given within the problem and in which it consists' (Simondon 2016: 156). But as Daniela Voss, in her consideration of the role of problems in Simondon's works, states, 'there is not really a generality to problems, much more they differentiate the individuation of non-living and living beings, and attribute a degree of indeterminacy in particular to psychosocial beings' (Voss 2018: 109). Problems gain traction through transductive operations, 'by which a structure appears in the domain of a problematic, that is, as that which provides the resolution of the posed problems' (Simondon 2009: 11). This solution is never predetermined, but has required an act of invention to be established, for the creation of a new passage between alien structures and potential energies to be actualised. Furthermore, this processural immanence implies the possibility of ethics, too, which for Simondon is expressed through the valuation of acts in their capacity for transductions. From this perspective, '[e]thics is nothing other than the affirmation of the inventions of life in all its forms, the setting into resonance of their differences, the reactivation of the openness of the pre-individual and the creation of new solutions to tensions, which generate new forms of living', writes Elisabeth Grosz in her concise chapter on Simondon (2017: 206-7).

It follows that individuation can not be known in the common sense, as CELIA LURY commences her contribution to this volume, because the knowing subject itself individuates with the problematic. The individuation of the problematic is the methodological concern Lury develops. And as a transductive operation that is inseparable from ontogenesis itself, any methodology of individuation then is nothing to select abstractly and to apply as if it was an unconstrained choice, but becomes operational itself: a constraint constrains itself as it is constituted in the very act. Lury refers accordingly to a 'compositional methodology' to signify this procedural character and to address the individuation of a problematic of the contemporary. Contemporary here is a term described by Paul Rabinow: 'The contemporary is a moving ratio of modernity, moving through the recent past and near future in a (non-linear) space that gauges modernity as an ethos already becoming historical' (2009: 2). A problematic of the contemporary is situated in that ratio which lets modernity emerge as it produces its history, and this ratio is the sole site of its actuality. Lury calls the environment of this individuation 'epistemic infrastructures', supporting becomings with materials of any kind, without being self-contained themselves, in an epistemic process that develops relations of knowledge to truth in the first place. As an example, Lury explores the implications of infrastructuring in urban spaces as real-time instrumentation in the form of sensed digital data that adds to such potentialities of individuation. Compositional methodology is thus concerned with uneven, nonlinear temporalities spurred by a plethora of epistemic infrastructurations and invests in the transitivity of methods, their transductivity for the grounding of new structurations. The aim is to test interdisciplinary methods for their compositional capacities towards problems, as a composite and compositional at once. Lury provides compositional examples from research concerning this auto-spatialisation instigated with methods that at the same time enter into the relation as they form it. For Lury, the contemporary concept of rendition, with its broad meanings, contributes to affective, moral and political outcomes as it negotiates the tension between an auto-as-autonomy and an auto-as-automatism in the auto-spatialisation instigated. Various styles of reasoning (induction, transduction, deduction) commit to various aspects of rendition, as do the multiplications of contexts. Her contribution in many ways complicates the polemics against 'the moderns', which have become rather fashionable in recent years, as it provides a problematisation of the relation between knowledge and truth that establishes an indetermination and thus retains potentials.

References

Althusser, Louis (1969): For Marx, London: Verso.
Amoore, Louise (2013): The Politics of Possibility: Risk and Security beyond Probability, Durham, NC: Duke University Press. https://doi.org/10.1215/9780822377269
Bachelard, Gaston (1949 [1966]): Le Rationalisme Appliqué, Paris: PUF.
Bachelard, Gaston (2012): 'Corrationalism and the Problematic.' In: Radical Philosophy 173, pp. 27-32.
Biagioli, Mario (1999): The Science Studies Reader, New York: Routledge.
Biagioli, Mario (2001): 'From Difference to Blackboxing: French Theory versus Science Studies' Metaphysics of Presence.' In: Sylvère Lotringer/Sande Cohen (eds.), French Theory in America, New York: Routledge, pp. 271-87.
Bianco, Giuseppe (2018): 'The Misadventures of the "Problem" in "Philosophy": From Kant to Deleuze.' In: Angelaki 23/2, pp. 8-30. https://doi.org/10.1080/0969725X.2018.1451459
Bowden, Sean (2018): 'An Anti-Positivist Conception of Problems: Deleuze, Bergson and the French Epistemological Tradition.' In: Angelaki 23/2, pp. 45-63. https://doi.org/10.1080/0969725X.2018.1451461

Carnap, Rudolf (2005 [1928]): Scheinprobleme in der Philosophie und andere metaphysikkritische Schriften, Hamburg: Meiner Felix. https://doi.org/10.28937/978-3-7873-2385-2

Cassou-Noguès, Pierre (2018): 'Cavaillès, Mathematical Problems and Questions.' In: Angelaki 23/2, pp. 64-78. https://doi.org/10.1080/0969725X.2018.1451463

Cavaillès, Jean/Canguilhem, Georges (1994): Œuvres Complètes de Philosophie des Sciences, Paris: Hermann.

Charbonnier, Pierre/Salmon, Gildas/Skafish, Peter (Eds.) (2016): Comparative Metaphysics: Ontology After Anthropology, London: Rowman & Littlefield International.

Collett, Guillaume (ed.) (2019): Deleuze, Guattari, and the Problem of Transdisciplinarity, London: Bloomsbury Publishing.

Debaise, Didier (2017): Nature as Event: The Lure of the Possible, Durham, NC: Duke University Press. https://doi.org/10.1215/9780822372424

Debaise, Didier (2018): 'The Minoritarian Powers of Thought: Thinking beyond Stupidity with Isabelle Stengers.' SubStance 47/1, pp. 17-28.

Deleuze, Gilles (1994 [1968]): Difference and Repetition, New York: Columbia University Press.

Deleuze, Gilles (1990 [1969]): The Logic of Sense, New York: Columbia University Press.

Dencik, Lina/Hintz, Arne/Redden, Joanna/Treré, Emiliano (2019): 'Exploring Data Justice: Conceptions, Applications and Directions.' In: Information, Communication & Society 22/7, pp. 873-81. https://doi.org/10.1080/1369118X.2019.1606268

D'Ignazio, Catherine/Klein, Lauren F. (2020): Data Feminism, Cambridge, MA: MIT Press. https://doi.org/10.7551/mitpress/11805.001.0001

Duffy, Simon B. (2018): 'Lautman on Problems as the Conditions of Existence of Solutions.' Angelaki 23/2, pp. 79-93. https://doi.org/10.1080/0969725X.2018.1451469

During, Elie (2004): '"A History of Problems": Bergson and the French Epistemological Tradition.' In: Journal of the British Society for Phenomenology 35/1, pp. 4-23. https://doi.org/10.1080/00071773.2004.11007419

Erdur, Onur (2018): Die epistemologischen Jahre: Philosophie und Biologie in Frankreich, 1960-1980, Zürich: Chronos.

Foucault, Michel/Rabinow, Paul (1984): 'Polemics, Politics, and Problematizations: An Interview with Michel Foucault.' In: The Foucault Reader, New York: Pantheon Books, pp. 381-397.

Gilson, Erinn Cunniff (2014): 'Ethics and the ontology of freedom: problematization and responsiveness in Foucault and Deleuze.' In: Foucault Studies 17, pp. 76-98. https://doi.org/10.22439/fs.v0i17.4254

Grosz, Elisabeth A. (2017): The Incorporeal: Ontology, Ethics, and the Limits of Materialism, New York: Columbia University Press. https://doi.org/10.7312/gros18162

Guattari, Félix (1995): Chaosmosis: An Ethico-Aesthetic Paradigm, Bloomington: Indiana University Press.

Haraway, Donna Jeanne (1988): 'Situated Knowledges: The Science Question in Feminism and the Privilege of Partial Perspective.' In: Feminist Studies 14/3, pp. 575-99. https://doi.org/10.2307/3178066

Haraway, Donna Jeanne (2004 [1985]): 'A Manifest for Cyborgs: Science, Technology, and Socialist Feminism in the 1980s.' In: The Haraway Reader, New York: Routledge, pp. 7-45.

Haraway, Donna Jeanne (2008): When Species Meet, Minneapolis: University of Minnesota Press.

Hirsch Hadorn, Gertrude et al. (2008): Handbook of Transdisciplinary Research, Dordrecht: Springer.

Kelly, Mark G.E. (2018): 'Problematizing the Problematic: Foucault and Althusser.' In: Angelaki 23/2, pp. 155-69. https://doi.org/10.1080/0969725X.2018.1451528

Klein, Julie Thompson (2014): 'Discourses of Transdisciplinarity: Looking Back to the Future.' In: Futures 63/1: pp. 68-74. https://doi.org/10.1016/j.futures.2014.08.008

Kuhn, Thomas S. (1996 [1962]): The Structure of Scientific Revolutions, Chicago: University Of Chicago Press.

Lautman, Albert/Duffy, Simon B. (2010): Mathematics, Ideas, and the Physical Real, New York: Continuum.

Lecourt, Dominique (ed.) (2006): Dictionnaire d'Histoire et Philosophie des Sciences, Paris: PUF.

Macherey, Pierre (1998): In a Materialist Way: Selected Essays, London and New York: Verso.

Maniglier, Patrice (2012): 'What Is a Problematic?' In: Radical Philosophy 173, pp. 21-3.

Maniglier, Patrice (2019): 'Problem and Structure: Bachelard, Deleuze and Transdisciplinarity.' In: Theory, Culture & Society, online. https://doi.org/10.1177/0263276419878245

Martin, Jacques (2020): L'individu chez Hegel, ed. by Jean-Baptiste Vuillerod, Lyon: ENS éditions. https://doi.org/10.4000/books.enseditions.14784

Massumi, Brian (2009): "Technical Mentality" revisited: Brian Massumi on Gilbert Simondon.' Interview with Arne De Boever, Alex Murray and Jon Roffe. In: Parrhesia 7, pp. 36-45.

Mittelstraß, Jürgen et al. (eds.) (2005-2016): Enzyklopädie Philosophie und Wissenschaftstheorie, 6 Volumes, Stuttgart: Metzler. https://doi.org/10.1007/978-3-476-00134-4

Osborne, Thomas (2003): 'What Is a Problem?' In: History of the Human Sciences 16/4, pp. 1-17. https://doi.org/10.1177/0952695103164001

Osborne, Peter (2015): 'Problematizing Disciplinarity, Transdisciplinary Problematics.' In: Theory, Culture & Society, 32/5-6, pp. 3-35. https://doi.org/10.1177/0263276415592245

Pignarre, Philippe/Stengers, Isabelle (2011): Capitalist Sorcery: Breaking the Spell. Basingstoke/New York: Palgrave Macmillan.

Rabinow, Paul (2009): Marking Time: On the Anthropology of the Contemporary, Princeton: Princeton University Press. https://doi.org/10.1515/9781400827992

Savransky, Martin (2016): The Adventure of Relevance, London: Palgrave Macmillan UK. https://doi.org/10.1057/978-1-137-57146-5

Savransky, Martin (2019): 'The Pluralistic Problematic: William James and the Pragmatics of the Pluriverse.' In: Theory, Culture & Society, online July 15th. https://doi.org/10.1177/0263276419848030

Serres Michel, and Nayla Farouki (eds.) (1997): Le Trésor. Dictionnaires des Sciences, Paris: Flammarion 1997.

Simondon, Gilbert (2005): L'individuation à la Lumière des Notions de Forme et d'Information, Grenoble: Millon.

Simondon, Gilbert (2009): 'The Position of the Problem of Ontogenesis.' In: Parrhesia 7/1, pp. 4-16.

Simondon, Gilbert (2016 [1958]): On the Mode of Existence of Technical Objects, Minneapolis: Univocal.

Souriau, Étienne (2015 [1943]): The Different Modes of Existence: Followed by, Of the Mode of Existence of the Work to Be Made. Minneapolis: Univocal.

Stengers, Isabelle (2009): 'William James: An Ethics of Thought?' In: Radical Philosophy 157, pp. 9-19.
Stengers, Isabelle (2015): In Catastrophic Times – Resisting the Coming Barbarism, London: Open Humanities Press.
Stengers, Isabelle/Debaise, Didier (2017): 'Towards a Speculative Pragmatism.' In: Parse 7, pp. 12-9.
Stengers, Isabelle (2019) 'Putting Problematization to the Test of Our Present.' In: Theory, Culture & Society, online July 15th. https://doi.org/10.1177/0263276419848061
Strathern, Marilyn (1988): The Gender of the Gift: Problems with Women and Problems with Society in Melanesia, Berkeley: University of California Press. https://doi.org/10.1525/california/9780520064232.001.0001
Viveiros de Castro, Eduardo (2014): Cannibal Metaphysics, Minneapolis: Univocal.
Voss, Daniela (2018): 'Simondon on the Notion of the Problem: A Genetic Schema of Individuation.' Angelaki 23/2, pp. 94-112. https://doi.org/10.1080/0969725X.2018.1451471

The *Problems of Modern Societies* – Epistemic Design around 1970

Isabell Schrickel

Historians characterise the years around 1970 as a period of multiple contradictions in modern societies, as the beginning of an era of fundamental transformations and sea-change – with long-term effects on societal, political and cultural developments in many parts of the world (Brick 2000; Judt 2005; Agar 2008; Suri 2009; Wirsching et al. 2011). The post-World War II economic expansion, the 'long boom', the 'trente glorieuses' or 'Wirtschaftswunder', which was based on cheap energy, enormous investments in science, technology and infrastructure, and productivity gains in many sectors, had reached critical limits – physical, ecological and ethical. High-modernist confidence, faith and optimism seemed exhausted and clashed – and sometimes coalesced – with waves of late- or postmodernist scepticism and doubts, but also with the activism and counterculture that had swollen throughout the 1960s. Some have identified the period as the beginning of our era due to emerging leitmotifs still relevant today, but also as an age of fracture, the seedbed of future crisis, and a landslide into uncertainty and ambiguity (Hobsbawm 1994; Jarausch 2008; Rodgers 2011; Ferguson et al. 2011). Deeply embedded convictions and beliefs of the preceding heyday of the Cold War became fragile, political conduct was questioned, and the meaning of distinctly modern key concepts such as progress and growth became less evident (Philp 2007: 169-213; Seefried 2015b; Andersson 2018: 122-150). Long-standing thought patterns of historical development – such as Marxism or modernisation theory – were challenged both on an ideological level and on the level of theoretical structure, as they were perceived as static, retrogressive, teleological and dichotomous. The idea that future developments and innovations in modern societies proceeded in progressive stages that were predicated on earlier ones and could thus be predicted seemed increasingly outdated. In the case of modernisation theory, such epistemological doubts matched with the growing qualms about the normativity of the West as a model for a completed and stable modernity (Gilman 2003: 203-276). And in the East the 'actually existing' state socialism entered a state of crisis as it missed launching the next phase of societal development. As an alternative to these competing models of political system development a new wave of interdisciplinary systems thinking gained momentum, imbued with motifs of complexity and interdependence, and focusing on the historical dependencies of societies, on new constellations of increasingly numerous actors in politics, the future as a potentially open-ended horizon of change, and global environmental interconnectedness and boundedness (Taylor 2001; Leendertz/Meteling 2016). In this context, the idea of the environment became a central discursive category at the time – particularly on an abstract system-theoretical level – as open systems are conceived as being reciprocally situated in a dynamic environment, but also on a practical level with regard to specific environmental problems receiving widespread public attention (Warde et al. 2018; Sprenger 2019). Intellectuals engaged with the complex dynamics of political change and the temporality of history as such in new ways. On the one hand, the long-term view of global change came into focus and increasingly sophisticated future projections, scenarios and imaginations gained new forms of agency on the present (Andersson/Rindzevičiūtė 2015). On the other hand, it was not only the nature and temporality of the future, but also the idea of history that changed, as the plural historicities of societies, knowledges and times as such were discovered and theorised in the humanities and the social sciences (Lorenz/Bevernage 2013; Esposito 2017). Consequently, these intellectual debates, conceptual innovations and semantic shifts indicate that the years around 1970 also need to be understood as an epistemic turning point (Leendertz/Meteling 2016; Rohde 2017; Heymann 2017). Following up from these observations, this paper seeks to bring together several mutually enforcing political, cultural and intellectual trends of the period and trace some characteristics of this broader epistemic shift, which fostered new modes of knowledge production, established various new fields of research and promoted new contexts for scientific collaboration: first, the political context of an abating Cold War and the development of more complex international relations; second, the rise of social movements that mobilised novel forms of expertise and critique; and third, an epistemological revolution that opened up new topics for inquiry, introduced new methodologies, forms of reflexivity and frames of analysis, and that made provisions for new roles of science. The notion of the 'problem' – from the *problematique* to the *wicked problem* – which occurs strikingly often in the period in public debates as well as in endeavours in the fields of systems science, future studies and theories of planning and design, will be analysed in its capacity to provide an 'epistemic design' for situations that call for a change, a transformation, specific interventions in a present state to actively envision a potential future.[1] This paper seeks to discuss the years around 1970 as a moment of transition in the history of modern societies by looking at how the notion of the problem became an operational term around which new modes of knowledge production, fields of research and interdisciplinary collaborations have emerged, and to trace aspects of the inventive epistemology of the 'problematic' as thematised in this volume in these contexts.

The historical context: an abating Cold War, the rise of social movements and new roles for science

During the 1960s, the key features of the Cold War – the arms race, the binary logic of the US-Soviet geostrategic rivalry and the ideological battle over whether authoritarian communism or liberal capitalism represented the preferable form of modern political economy – began to abate in intensity. Nuclear weapons appeared more and more as an 'ideology killer' because the physical stakes they produced seemed higher than the ideological ones, and resourceful, military-based conflict theories and the doctrine of deterrence didn't seem to make the world a predictably safer place any more (Gilman 2016). The 13 days of the Cuban Missile Crisis of October 1962, when there 'was a higher probability that more human lives would end suddenly than ever before in history' (Allison 1969: 689) reverberated for a long time and gave rise to strategic reconsideration. One lesson that came out of it was the extent to which the adversaries had failed to think similarly going into it. What had appeared to be 'rational' behaviour in Moscow had come across as

---

1 While Scopus tracks a constant increase in the occurrence of the term 'problem' in article titles between the years 1950 and 2000, the increase between 1970 (5,381 article titles) and 1974 (8,338 article titles) is at 55 per cent by far the largest over the whole period.

dangerously 'irrational' behaviour in Washington, and vice versa. Hence, in practice there were sharp limits to the unilateral rationalisation and solution of critical situations through game-theoretical exercises or rational choice frameworks promoted by the so-called defence intellectuals. The 1960s held a number of developments – the Sino-Soviet split, Ostpolitik, decolonisation and the political rise of the 'Third World'[2] – that further complicated the dualistic framing and opened up opportunities for – or perhaps exerted pressure on – other forms and layers of international relations and diplomacies, multipolarity and the forging of complex interdependence that would mark the next period of international history.

Those were also the years in which social movements made themselves heard globally and became highly visible features of politics and culture beyond the turn of the decade. In many ways they reflected the larger implications of the aforementioned political shifts. Discord in the communist world prompted ideological crisis, while Western civil societies began to question their own values, integrity and righteousness. New social movements – pacifism, feminism, environmentalism – drew attention towards various issues previously neglected or suppressed, from democratic participation to civil rights and problems of the environment. Most emblematically, the turmoil culminated in the events of 1968 – the worldwide protests against the war in Vietnam, the student movements in France, West Germany, Poland, Japan, the USA, Mexico, Tunisia and other places, the Prague spring liberalisation in Czechoslovakia, and the civil rights and environmental movements (Suri 2009). An essential characteristic of these movements was the appropriation of new forms of public expression and communication, educational practices and global solidarity. In the light of détente, looming environmental crisis, the computer and communication revolution, global protest movements and educational revolts, heretofore unquestioned modes of governance, ideas of order and control, patterns of thought, policy cultures and epistemologies were challenged in various ways. Calls for greater political participation and the developments of fields of political engagement at communal and national levels – but also within the United Nations – indicate that the shifts of the years around 1970 opened up paths into less hierarchical societal developments and polycentric orders (Cox 1981; Christiansen/Scarlett 2013). It

---

2   During each year from 1960 through 1970 an average of three states gained their sovereignty, most of the new states being in Africa.

became apparent that the sources of power in society and international relations had begun to diversify at an unprecedented scale. Various supranational organisations emerged as influential political actors, introducing a greater diversity of goals pursued internationally and producing greater complexity in the modes of interaction and the institutions within which action takes place (Cox/Jacobson 1973). The future of modern, industrialised societies became a major concern, and knowledge was increasingly conceptualised as an important resource in these cultural and political transformations.

Accordingly, these political shifts had a distinctive impact on science, both as a model of inquiry and as a modern tool for progress as the relationships between science and the questions it sought to answer were fundamentally transformed. We can understand the intense debates on epistemology and inter- and transdisciplinarity at the time as manifestations of these transformations. The contemporary struggles over epistemology in France, for example, are evaluated today as instances of an important mutual exchange between the sciences, philosophy and society, providing novel techniques and tools for argumentation, thought and action and a specific mode to reflect on the role of science in society (Erdur 2018). Scholars from different fields were also increasingly concerned about how to bring science closer to real-world problems and find solutions to pressing social, political and environmental issues, as well as with the future of universities and education as such. These issues were often thematised as problems of disciplinary boundaries and how to transgress them.[3] Varieties of (post-)structuralism, systems theory and cybernetics, with their transdisciplinary conceptualisations, provided important frames of reference for engagement with a less universalist and more situated understanding of science as a social process and vital resource for problematisation and cultural change.[4]

---

3 The notion of 'transdisciplinarity' was introduced at a meeting on interdisciplinarity in universities held at the University of Nice and jointly sponsored by the Organisation of Economic Co-operation and Development (OECD) and the French Ministry of Education. The Swiss psychologist Jean Piaget, the French mathematician André Lichnerowicz, the Austrian astrophysicist Erich Jantsch and the British media historian Asa Briggs were among the participants. See Apostel et al. 1972; and for historical and analytical accounts on inter- and transdisciplinarity see Barry/Born 2013; Schaffer 2013; Klein 2014.

4 Osborne (2015: 14-15) claims that 'the disciplinarily disruptive and transformative forces' of the great books from the European humanities – many of which this volume discusses – provided the ground for inter- and transdisciplinarity movements.

The promotion of new constellations between scientific communities and increasingly transnational publics, and of unprecedented forms of scientific collaboration within newly formed international institutions, can be understood in this context. Another manifestation of the transformation of the intellectual landscape at the time was the diversification of the scientific persona: action intellectuals, technocrats, institutional entrepreneurs, the radical science movement, scientific activists, engaged intellectuals, science policy experts – all began to interact with one another, and many emerged from hidden committees into the public eye (Agar 2008; Shapin 2009; see also White 1967; Nelkin 1979).

The epistemological revolution

To a large extent accounts of the history of science during the Cold War have focused on instances that reveal how the Cold War drove scientists and science policy ideologically and 'distorted' the evolution of science. The question of how emerging ideas about the global environment and an interconnected and interdependent world system, eroding epistemic certainties and shifting values, might have challenged binary world-views, the self-evidence of particular ideologies, models of society or the possibility of an objective science at all, is only recently being posed. The same applies to the overall plausibility of national security imperatives or the competition and rivalry between the systems as comprehensive frames of reference for developments in science policy during the Cold War. There is a growing demand and interest in contextualising emerging fields of research within longer-term intellectual trends, changing research infrastructures and innovative institutional environments (Engerman 2010).[5] Additionally, the developments outlined so far indicate that the years around 1970 marked a transition in the post-World War II history of science. Besides the aforementioned greater diversity among the scientific personnel, emerging inter- and transdisci-

---

5 Histories of interdisciplinary fields such as cybernetics, futures studies, policy sciences, science studies, peace studies or environmental sciences have provided instructive examples that complicated overly simplistic and exploitative ideas of the relationship between power and knowledge (Gestwa/Rohdewald 2009; Thomas 2015; Seefried 2015a; Rindzevičiūtė 2016; Gilman 2016; Graf 2017; Rohde 2017; Andersson 2018).

plinary perspectives and movements, new forms of international collaboration, and the processes of renewal many existing disciplines and institutions went through, we should acknowledge the fact that scientists and intellectuals often played an active role in imagining a global, non-apocalyptic post-Cold War world. They posted different sets of objectives for human development and thematised issues in ways that challenged some of the operating conditions of modern societies and the status quo of international relations. They sought to influence future developments by creating new institutional set-ups or through specific problematisations around which new fields of research emerged. A distinctive characteristic of the social and human sciences around 1970 was an orientation towards greater reflexivity, autonomy and new forms of intellectual engagement, which opened up the 'closed world' of Cold War science (Edwards 1996). Many fields went through what historian Peter Novick described in 1988 as an 'epistemological revolution', that is, a break with the prevalent post-World War II model of social and human sciences inquiry. Up to the 1960s the philosophy of the social and human sciences rested on the belief that they were, in the main, value-free disciplines and 'an extension of the dominant positivist and empiricist philosophy of the natural sciences' (Novick 1988: 546). One of the common visions between such diverse enterprises as rational choice theory, structural-functional sociology, information theory or operations research, among others, was the study of 'systems' behaviours, the causal chains within systems of variables whose interrelations could be formally stated and studied in technical terms (Isaac 2012a: 9; Heyck 2015). Joel Isaac has discussed the problem of 'epistemic design' in this context, that is, how in post-World War II American social science empirical knowledge was constituted within the framework of a scientific theory through particular arrangements of data and techniques of representation (models, diagrams, tables) 'so as to make them undergird the theoretical claims about the social world they wished to make' (Isaac 2012b: 80). And for various reasons that are obviously connected to the early Cold war setting this world was rendered knowable, predictable and controllable. Such arrangements served the purpose to claim an 'objectivism' and 'scientism' for the social sciences in the post-World War II context. In the years around 1970 however, some of these objectivist claims and assumptions came to be undermined. According to Novick, 'in one field after another distinctions between fact and value and between theory and observation were called into question' as the 'notion of a determinate and unitary truth about the physical or social world' came to be seen by a growing number of scholars as a 'chimaera'. For him the 'meaning of *progress* in science and scholarship became problematic'. While Novick suggested that it was 'for the most part "strictly academic" considerations which initiated debates, and contributed the categories in which heterodox views were advanced', so far this paper has delivered some reasons to rethink the 'epistemological revolution' as a primarily 'academic' endeavour (Novick 1988: 523, 546). It should rather be understood as a complex and co-evolutionary process in which the procedures of science interacted more intensely with the social, political, technological and intellectual environment than before and that challenged the prevailing positivist problem-solving mentality in many disciplines. Novick's monumental effort to examine the 'objectivity question' in the historical profession is in itself a result of these broader transformations.

In order to support his argument on the epistemological revolution as a mostly academic endeavour, Novick draws largely on the cross-disciplinary circulation and adoption of Thomas Kuhn's 1962 *The Structure of Scientific Revolutions*. Recent accounts of Kuhn's work, however, have positioned him in an elitist and prestigious Harvard context during the Cold War, where he was challenged to formulate a theory of science that represented a 'distinctive response to the pressing problems of epistemology and society' at the time (Isaac 2012a: 193).[6] Kuhn's *Structure* was indeed an important point of reference and provided a conceptual framework within which to discuss the practical and evolving nature of scientific inquiry for a wide range of actors at the time – including the sociologists, philosophers, systems thinkers, natural scientists, international relations scholars, policy advisors and institutional entrepreneurs mentioned in this paper.[7] The book introduced the broader academic world to a non-teleological, evolutionary, historical view of scientific development as a sequence of incommensurable but nevertheless internally consistent periods of 'normal science', operating within

---

6 For a comprehensive historical contextualisation of Kuhn's *Structure* see Fuller 2000; Reisch 2016.

7 Novick, just like many other American intellectual historians, largely draws on Kuhn for an early historical and situated conceptualisation of scientific development. But there is a much richer French historical-epistemological tradition, which Novick mentioned only briefly. Authors like Gaston Bachelard, Georges Canguilhem, Michel Foucault or Gilles Deleuze have much to say about inventiveness and the sources of the new in scientific inquiry (cf. Rheinberger 2010).

the conceptual framework of a 'paradigm'. From time to time 'puzzle-solving' normal science is turned upside down, 'gestalt switches' occur, and a choice has to be made 'between incompatible modes of community life' – a paradigm shift takes place (Kuhn 1962: 94, 117). This view of normal science inverted the image of the scientist in an interesting way: in its suggestion that 'dogma was the precondition, not the antithesis, of scientific advance, and in its corollary – the "normal" scientist as tradition-bound puzzle-solver, rather than bold adventurer – it fundamentally contradicted the orthodox Promethean image' of the scientist (Novick 1988: 529). While Kuhn himself was probably more concerned with the question of how paradigms are actually stabilised and ultimately embraced a pedagogical theory of science[8], the epistemological adventurers of the late 1960s – systems thinkers, policy experts, futurologists, engaged intellectuals, institutional innovators – embraced the revolutionary moment and strove for the establishment of new paradigmatic frameworks, perspectives that would help to open up the world for malleability and active intervention. These dynamics became visible in the processes of renewal many existing disciplines went through, but also in emerging fields of research that resonated with the social, political and environmental situation at the time. The notion of the *problem* – ubiquitously applied and famously pored over at the time from debates at students' kitchen tables over the Club of Rome's talk on the 'world problematique' (Özbekhan 1970) to the discovery of 'wicked problems' (Churchman 1967; Rittel/Webber 1973) – is understood here as a marker of this revolutionary atmosphere.

Branches of contemporary social science started to question their commitment to objectivity and related ideals as well as the preference for quantitative analysis as opposed to historical or other 'soft' forms of social research at the time. The idea of value-neutrality was rejected and a reorientation of the social sciences toward normative analysis was discussed (Solovey 2001). The crucial importance of values as *dynamic* factors in societal processes was re-discovered during the 1970s, and studies on value shifts – for instance the orientation towards post-materialism in Western societies – became a central field of research (Galtung 1970; Schumacher 1973; Inglehart 1977). Furthermore,

---

8 According to Kuhn paradigms are stabilised through exemplary scientific achievements or model experiments, theoretical and ontological assumptions (amounting to a disciplinary *Weltanschauung*), professional training, methodology, instrumentation and research agenda; (Kuhn 1962: 35-42).

the rise of new media, the rapid spread of computers and the unprecedented technological possibilities of communicating across the globe gave reason to study the structural transformation of the public sphere and cultural dynamics in relation to communication and media (Habermas 1962; McLuhan 1964). Influential media theories such as the agenda-setting theory, the knowledge gap hypothesis and framing theories were developed in the years around 1970 (Tichenor et al. 1970; McCombs/Shaw 1972; Goffman 1974). In addition to these new or strengthened fields of social science several historians marked that at the peak time of the space age the unprecedented and iconic global environmental images of the blue marble shaped a newly global environmental consciousness, resulting in an 'ecological revolution' (Radkau 2014; Seefried 2015b). Consequently, the 'vast machine' of the environmental sciences emerged as a global knowledge infrastructure, a large-scale sociotechnical system collecting environmental data and modelling and projecting planetary processes (Jasanoff 2001; Edwards 2010; Cosgrove 2001). In particular, climate science changed at the time, from a mostly descriptive and heuristic research programme into an interdisciplinary programme in which interactions between humans and the Earth system were studied and predicted via computer simulations (Heymann 2009). Research on planet Earth and its systemic properties and interactions gave rise also to new ideas on the politics and economics of the planet, its limited resources, boundaries and how to legislate its use (Boulding 1966; Hardin 1968; Georgescu-Roegen 1971; Daly 1974).

The years between 1964 and 1973 were also the high point of futures research. Distinct from the 1950s and early 1960s futurology that studied foreseeable laws of social development with a positivist mindset, futures studies' point of departure was the 'historically specific understanding that the present was a far from stable structure' that cannot be predicted in a positivist sense. It was in several ways a form of 'counter-expertise to the project of Cold War prediction', with strong links to international activist movements, as Jenny Andersson has recently argued (Andersson 2018: 3, 47; see also Seefried 2015a). Predicting the future based on a status quo was considered to primarily serve the elites, as it perpetuated this status. Thus, new forms of futures techniques were developed that allowed people to actively shape and develop alternative futures. Approaches such as the Argentinian philosopher Paulo Freire's critical pedagogy aimed to help Latin America's poor people reach self-consciousness and inspired a whole generation of European social workers from the late 1960s on (Freire 1970; Andersson 2018: 151-183).

At the same time in Europe, Robert Jungk developed the future workshop technique (*Zukunftswerkstatt*) in the context of specific controversial public policy issues, which was conceived as a process lasting several days that aimed to unleash the social imagination of the (non-expert) participants by using the tools of radical and dialectical deconstruction and psychotherapy (Jungk 1987; Andersson 2018, 151-183). Compared with methods of governmental future planning, such as systems analysis and operations research, these alternative and activist futures techniques held utopian and emancipatory aspirations, as they aimed at an open process in which problems were to be constructed and a set of objectives for social and personal development was to be established. At the same time, however, it was a period in which the application of systems analysis and operations research was widely expanded: initially developed in military contexts to improve defence efficiency and to underscore the rationality of decision-making (Hughes/Hughes 2000), systems thinking became more important in engineering and in the hard sciences, as they provided powerful tools for the control, optimisation and prediction of complex systems. During the 1960s, systems analysis was increasingly expanded to public policy, social issues, and urban and environmental planning (Jardini 1996; Light 2004). The design and results of such policy processes were widely criticised at the time, for example by the sociologist Ida Hoos from Berkeley and Robert Boguslaw from the RAND Corporation. Hoos emphasised that the current practice of systems analysis techniques such as cost/benefit ratios or programme budgeting led to technocratic forms of governance, because policies were often crafted by outside consultants and experts 'armed with solutions and in search of problems', specialised in managerial efficiency, but not political problem consideration. Hoos saw growing incidences of 'government-by-contract' situations that removed responsibility for the decisions made from public officials (Hoos 1972: 86, 243). Boguslaw argued in a similar manner that systems analysts built models that, as a result of the experts' attempts to be value neutral and objective, reified the values of the ruling elite and implicitly supported the status quo (Boguslaw 1965). Towards the end of the 1960s, Boguslaw left RAND and took a position at the progressive sociology department of Washington University in St. Louis, where he developed – in the eyes of his former colleagues – a rather obscure interest for French existentialist philosophy, referring to authors such as Jean-Paul Sartre, Louis Althusser and Jacques Ellul in his evaluations of solutions to the problems of modern society proposed by contemporary social science and other expert panels (Rohde 2013: 53-59). However, such profound criticism and debates led to a more reflexive understanding, experimental opening up and further modification of systems methods. Boguslaw, Hoos and others certainly placed high expectations on the potential of systems analysis as a 'social technology' (c.f. Helmer et al. 1966) and as a 'phenomenon fraught with social significance, perhaps all the more because it is characterized by contradictions, internal and external' (Hoos 1972: 241). Thus, more situated frameworks for systems approaches have been developed in the context of specific problem areas such as environmental assessment, urban planning, public policy and other design challenges where systems techniques have been further developed into a rich and interdisciplinary field of increasingly reflexive, more situated and post-positivist policy, decision and management sciences (see for instance Anderson 1968; Rittel/Webber 1973; Holling 1978). A new generation of scholars and practitioners in different fields questioned previous hyperbolic notions of controllability, knowability and rationality. Instead, they were increasingly interested in the 'epistemology mediating between organizations and the welter of experience surrounding them' (Dery 1984: 118), in more adaptive and participatory processes of planning, in shaping more carefully the processes of constructing a problem and more generally the role of knowledge production in public policy in the light of debates about increased social 'complexity' (Leendertz 2019). Many scholars began to understand systems approaches as a means to construct and invent problems in the first place – not to solve prescribed ones. A contemporary observer commented on the conjuncture of systems science as an ongoing 'social experiment' with an open outcome (Churchman 1978). The problem of epistemic design thus shifted at the time – or rather, it was regularly complemented by the idea that systems approaches can be applied to pose and construct problems so as to make them create images of change or suggest particular policy options.

Finally, the years around 1970 were also a foundational period for what was to become probably one of the most innovative and growing fields of reflexive and critical social and human science inquiry of the last third of the 20th century, that is, the various versions of studies of science – the philosophy, history and sociology of science and technology. The establishment of science studies at the time can be understood both as an effect of the aforementioned developments and debates and a symptom of an epistemic turning point. These studies conceptualised science and technology as essential drivers of social, political and environmental change and as constitutive for the future of modern societies. Early versions of meta-studies of science emerged from different contexts, for instance from debates on the ethical implications of advances in particular fields such as molecular biology or on the risks and side effects of new technology such as nuclear energy, but also from transnational debates about the impacts, governance and role of science and technology both in history and for the future of industrialised societies, as new global challenges began to emerge. These debates were, as Elena Aronova has shown, thoroughly embedded in the 'political economy, cultural anxieties, and ideological dimensions of the post-World War II social and political order'.[9] But at the same time many debates reveal that in those years an intellectual space opened up in which science as a complex cultural phenomenon was conceptualised and advanced in new ways. Based on experiences with the numerous roles international scientific cooperation had played in the normalisation of relations after World War II, science policy practitioners envisioned new areas of science diplomacy and international knowledge co-production. While historians have shown how early Cold War US-American science policy was embedded in the projection of American political and ideological interests in postwar continental Europe, and how cooperation served as a means of fostering a consensual or co-produced hegemony in order to consolidate a liberal, democratic, capitalist regime in Europe (Krige 2006; Doel/Wang 2001), the role and prospects of science in international relations began to change towards the end of the 1960s. Even though the world was divided, it was also perceived as being increasingly interdependent, meaning that the autonomy of nations was becoming limited by transnational flows of energy and goods, of money and ideas, and even of pollution and diseases – and scientific and technological collaboration were discussed not only in US foreign policy as ways of shaping a mutual understanding of the problems and concerns that modern societies had in common (Rosenau 1969). We find numerous examples for these changing relationships between science and power in an interdependent world, emerging collaborative organisations and also the debates concerning the political,

---

9 These studies had different names in different places: 'science studies' in the USA and UK, 'naukovedenie' in the Soviet Union, 'naukoznawstwo' in Poland, 'naukoznanie' in Bulgaria, 'natural dialectics' in China, 'Wissenschaftswissenschaft' in East Germany and in West Germany the 'Finalisierungsdebatte' has to be mentioned (Aronova 2012a; Leendertz 2013).

ethical and social implications of science and technology more generally. The perceived urgency to navigate the interdependencies and future paths of these relationships fostered novel forms of expertise, advanced scientific internationalism and promoted closer relationships between science and policy.

The future of modern societies

In these shifting political, cultural and epistemological terrains of the late 1960s, transnational debates began to emerge, in which deeper concerns about the future development of modern, industrialised societies in the context of rapidly changing techno-scientific environments, ecological limits, interdependent economies and shifting values came to the fore. At the core of these debates was the observation that as a result of the rise of technical and scientific expertise in modern societies, and its growing importance in the realm of public affairs, ideology was losing its revolutionary potential and organisational power. Neither Marxian theory of historical development nor the Western counterparts such as Rostow's modernisation theory seemed to provide convincing theoretical devices to confront the profound changes looming ahead. Both offered teleological ideas of societal progress as an essentially predictable process of social change (Gilman 2003; Andersson 2018: 49). These frameworks have been replaced by theories of 'convergence' and the 'post-industrial society', developed independently with different foci and nuances by diverse authors such as Raymond Aron, Daniel Bell, Alain Touraine, Peter Drucker, Pitirim Sorokin, Jan Tinbergen, Kenneth Galbraith, Samuel Huntington and Jacques Ellul. These authors envisioned that industrial economies around the world would be converging in terms of social, economic and political structure. Additionally, social and political thought would be converging on a remarkably widespread agreement over advanced societies' fundamental aims, and the focus would shift onto the *problem* of achieving such aims.[10] In light of the perceived challenges of peacekeeping at the time, the global population boom, sustenance, energy demand, over-

---

10 Drucker 1950; Ellul 1954; Sorokin 1960; Tinbergen 1961; Aron 1963; Brzezinski/Huntington 1964; Galbraith 1968; Touraine 1971; Bell 1973; on the political contexts of these debates see Aronova 2012b; Andersson 2018: 49-74.

exploitation, automation and the biological revolution, numerous authors began to theorise the temporality of historical dynamics, the governability of societal developments and the question of how to bring about desired change (Buchholz 1968; Huntington 1971; Luhmann 1972; c.f. Esposito 2017).

Closely related to these ideas, an emerging network of influential policy experts on both sides of the Atlantic was concerned with the impact of the 'scientific-technological revolution' (in the East), the 'technetronic era' or the 'post-industrial society' (in the West) as a revolutionary challenge to both paradigms of historical development. The Czech philosopher Radovan Richta, one of the minds behind the 'socialism with a human face' reform movement, for example, published *Civilization at the Crossroads* in 1966, in which he suggested that the revolution was not a matter of a future endpoint, but a continuous process in the present with an open horizon of change in a highly technological and science-based society that would require new forms of participation (Richta 1969; Sommer 2017; Andersson 2018: 128-129). In 1969 the Italian industrial consultant and founder of the Club of Rome, Aurelio Peccei, analysed in *The Chasm Ahead* the 'menacing technological gap that now separates the United States and Europe'. He described a 'world in convulsion' due to 'the unprecedented complex of explosive problems', 'technological acceleration' and 'exponential growths' in population, pollution, automation etc. Similarly to Richta, he perceived the present as a 'period of revolutionary and even metamorphic change' that required collaborative and interdisciplinary efforts in order to master the future (Peccei 1969: 104). In a similar vein, Zbigniew Brzezinski, éminence grise of US-American foreign policy, in his 1970 *Between Two Ages: America's Role in the Technetronic Era*, cautioned against a cognitive crisis between and within modern societies. He saw a 'threat of intellectual fragmentation, posed by the gap between the pace in the expansion of knowledge and the rate of its assimilation', which would 'raise a perplexing question concerning the prospects for mankind's intellectual unity. It has generally been assumed that the modern world, shaped increasingly by the industrial and urban revolutions, will become more homogeneous in its outlook. This may be so, but it could be the homogeneity of insecurity, of uncertainty, and of intellectual anarchy. The result, therefore, would not necessarily be a more stable environment.' (Brzezinski 1970: 23) On the other hand, Brzezinski also observed that an emerging 'planetary consciousness' and the 'availability of the means to cooperate globally' intensified the 'sense of proximity' and the 'sense of obligation to act' (Brzezinski 1970: 60).

These rhetorics of fractures, chasms and crossroads paved the way for greater international interest in similar systemic characteristics and outlooks of the future of modern societies, and they became a point of departure for critical debates on the common problems, direction and governability of these developments (Leendertz/Meteling 2016). Even though these engagements had been developed in quite different political contexts, they shared a sense of urgency and being present at a turning point in the history of modern societies – demanding new theoretical devices, interdisciplinary perspectives and methodologies. The novelty of the problems associated with the political and cultural shifts, the advances in science and technology making these problems discernible, global interdependence and the environmental crisis looming ahead were widely perceived as challenges for high modernist self-perceptions, approaches and expectations.

A post-positivist conception of problems

In the following I argue that the intellectual impact of this epistemological revolution and the transition towards post-positivism can be understood through the conjuncture and the changing status of the notion of the *problem* in intellectual endeavours at the time. Before I turn to exemplary cases, let me briefly recall the etymology. The word *problem* stems from the Greek verb *probállo* (to put forward, to propose, to outperform somebody) and the noun *próblema* (cliff, obstacle, question, exercise). From the meaning it becomes clear that problems can be spontaneously emergent in a situation or they can be purposefully posed to a counterpart. They can be concrete and tangible but also conceptual and abstract. The prefix *pro-* can be understood temporally as something that lies *ahead* and demands a reaction or a strategy. On the other hand, it points towards a strategy of empowerment or authorisation through which someone can pose a problem *for* someone else. It is both a prefix of priority in space or time, having a meaning of advancing or projecting forward or outward, and/or it indicates substitution. In a positivist understanding of science[11] the concept of the problem functions as some

---

11 Positivist conceptions of science are usually bound to a combination of the following accounts: *realism*: truths about the world are true regardless of what people think and there is a unique best description of any chosen aspect of the world; *demarcation*: there is a dis-

kind of placeholder for the time span needed to find the solution. Problems are obstacles to be removed, means to implement specific solutions, negative states of uncertainty, ignorance and methodological imperfection bound to dissipate with the solutions that scientific and technological progress yield. Consequently, traditions like logical positivism rejected the 'great questions': philosophical, metaphysical, vital and singular problems 'are in fact *pseudo-problems*, which are incapable of solution not because of their profundity but because they pose nothing to be solved' (Kaplan 1968). In a positivist understanding problems are conceptualised negatively in the sense that they are meant to disappear. They are chosen precisely because nothing will be left from them – just like a puzzle. In a post-positivist understanding, it is particularly this aspect that is different. Problems are actively constructed as matters of concern in order to intervene in the present and to create agency and images of change. They are devices to open up and assure us of some creative leeway and measure of control over an uncertain future. The particular ways problems are constructed imply how they function as political or social technologies.[12]

In the light of the aforementioned developments and transnational debates on the challenges of modern societies, many institutional entrepreneurs, policy advisors and systems thinkers engaged with this particular understanding of the problem – both on an institutional level and as an intellectual agenda but also in methodological and theoretical writings. They

---

tinction between scientific theories and other kinds of beliefs or knowledges; *axiomatisation*: the content of a theory is analysed as a given set of axioms from which the remaining content of the theory can be derived deductively as theorems; *reductionism*: phenomena can be described in terms of other simpler or more fundamental phenomena; the *unity* and *universality* of science: both hypothesis and working programme, this claim states that all scientific disciplines are part of the same endeavour and that less profound sciences are reducible to more profound ones; and the *cumulative* character of science, the Whiggish narrative of continuous scientific progress through an elimination of falsehoods by the discovery, verification and systematisation of empirical truths or facts (c.f. Mittelstraß et al. 2005-2016).

12 The concepts of 'social technology' and 'political technology' describe how forms of applied knowledge can be used in pragmatic and intentional ways to transform states or societies have been used by different authors with varied nuances throughout the 20th century, among them Thorstein Veblen, Karl Polanyi, Karl Popper, Olaf Helmer and Michel Foucault. For a critical review see Leibetseder 2011; for the Soviet debate see Rindzevičiūtė 2015.

deployed the notion in the sense of a post-positivist 'epistemic design'[13] in order to address the procedural criticality, complexity, uncertainty and openness of recent societal developments in modern societies and their increasingly critical relationships to a globalised environment as a continuous epistemic and political challenge that requires innovative institutions, and novel frameworks and approaches, but also to suggest particular policy options over others. The notion of the problem became a politically operational and performative concept, not a mere placeholder within a clearly defined scientific task. In many cases the ambition to construct problems in order to leverage systems into a different state became apparent. In that sense, problematisations provided an epistemic design that could be used to engage with the question of the future development of modern societies, a procedure that allowed the gathering and arranging of data, modelling issues, defining their aspects and boundaries, and deriving options for action. The particular ways in which problems became thematic in programmatic writings and institutional agendas can be understood as a conceptual turning point and an epistemological effort in this transitional period in the history of modernity.

In the French epistemological tradition, a similar understanding of the notion of the problem has been substantially theorised. Based on the philosophy of Henri Bergson, Gilles Deleuze, for instance, has provided a useful evaluation of the truth or falsity of problems. A true problem is one that is intrinsically productive, a kind of ongoing, groping and experimental process that forces the transformation of the subject's thought (During 2014). A false problem, by contrast, is one whose determination depends on something external, an extrinsic conditioning as opposed to an intrinsic genesis. Here the substituting, instrumental meaning of the word is addressed, which gives reason to carefully distinguish between problematisations that are chosen because they allow the implementation of extrinsic pre-defined measures and solutions and those that allow an intrinsic differential evolution of a problematic situation (Deleuze 1968; Bowden 2018). For an epistemological

---

13 On 'epistemic design' see Isaac (2012b, 80, 88), who defines epistemic design as the concern with the problem of 'how to arrange [the social scientist's] data so as to make them represent and undergird the theoretical claims about the social world they wished to make' and the challenge of how 'to arrange the data so that they could serve as instruments for investigating and perhaps even manipulating value systems in actually existing social systems'.

assessment of post-positivist scientific approaches in specific historical contexts, and for analysing how problematisations provide an epistemic design of specific matters of concern, this might be a helpful distinction to make. It helps to recognise critical approaches that are truly productive and transformative in character but also to identify cases where problems are used in rather instrumental terms, which also resembles the criticism voiced by Hoos and Boguslaw.

Researching through the problematique: institutions, issues and designs

Among the more radical adopters of the notion was the Armenian, Turkey-born, systems scientist Hasan Özbekhan, who had studied at the London School of Economics, then worked for RAND and the Systems Development Corporation in Santa Monica. He was the author of a number of influential writings in which he outlined theoretical approaches around the notion of the 'problematique' that circulated between the Organisation for Economic Co-operation and Development (OECD), RAND, the Club of Rome and the network of futurists from the late 1960s. Özbekhan argued that 'the all-pervasive analytic or positivistic methodologies' modern planning processes rely on 'failed to provide us with an ethos, a morality, ideals, institutions, a vision of man and of mankind and a politics which are in consonance with the way of life that has evolved as the expression of our success'. At the OECD Working Symposium in Bellagio on long-range forecasting and planning in 1968, with an illustrious group of participants such as Jay Forrester, Erich Jantsch, René Dubos, Stafford Beer, Aurelio Peccei and others, Özbekhan outlined a 'General Theory of Planning' in which he rejected the idea of value-free approaches. Recognising the current 'problematical situation' means that 'there exists a dissonance between the situation and the value system: [...] If planning is viewed as a problem-solving device, then the emphasis of action is to bring changes in the environment while leaving the value system untouched and thus to achieve consonance between the two. If planning is viewed as a continuous organisation of progress throughout the environment, then it becomes necessary to effect changes in the value system as well as in the environment to achieve consonance between the two.' (Özbekhan 1969: 152) Consequently, his first technical and methodological proposal for the Club of Rome was designed to 'cognize and investigate the all-pervasive *problematique* which is built into our situation, through some new leap of inventiveness'. Özbekhan's generalised problematique was a 'system-wide, interdependent, interactive and intersensitive [complex], immune to linear or sequential resolution, [and] ecosystemic in character' – it posed nothing solvable, but something 'that inheres in our situation'. In order to understand and intervene in the dynamics and to reach 'ecological balance' he suggested a 'coarse graining' of the complex problematique by identifying a set of 'continuous critical problems' covering the 'biological, physiological, physical, psychological, ethical, religious, technological, economic, political, national, international, communal, attitudinal, intellectual, institutional' aspects of it. A combination of conceptual and axiomatic work and a cybernetic system would allow study of the behaviours of different set-ups and exploration of both the 'logical, normal future (forecast)' and a 'normative future, imagined in the light of the value-base of ecological balance'. In that sense, the aim of Özbekhan's proposal was 'not research in the traditional sense but *invention*' (Özbekhan 1970).

The activities of the Club of Rome were organised for some time around Özbekhan's idea of a highly relational or interdependent 'world problematique' as the 'complex of problems [...that] occur to some degree in all societies; they contain technical, social, economic, and political elements'.[14] The call for inventiveness was heard by Jay Forrester at the Massachusetts Institute of Technology (MIT), who together with his team translated the structure of the problematique into a computer model, which the famous *Limits to Growth* study was based on. However, many aspects of Özbekhan's comprehensive, ambitious and reflexive proposal were neglected in Forrester's systems dynamics approach. The criticism *Limits to Growth* received, particularly

---

14 '[T]he complex of problems troubling men of all nations: poverty in the midst of plenty; degradation of the environment; loss of faith in institutions; uncontrolled urban spread; insecurity of employment; alienation of youth; rejection of traditional values; and inflation and other monetary and economic disruptions. [...] It is the predicament of mankind that man can perceive the problematique, yet, despite his considerable knowledge and skills, he does not understand the origins, significance, and interrelationships of its many components and thus is unable to devise effective responses. This failure occurs in large part because we continue to examine single items in the problematique without understanding that the whole is more than the sum of its parts, that change in one element means change in the others.' (Meadows et al. 1972: 10-11)

because of the methodologically closed approach, the lack of data, the computer fetishism and the doomsday mentality, became an iconic tale on the ethical and epistemological challenges of experimenting with high ambition in the realm of the problematique (Vieille Blanchard 2010).

At the same time, within the OECD, initially founded to coordinate economic policies among the Western nations and first and foremost seen as a technocratic institution implementing the paradigm of economic growth in the Cold War setting, an ecologically oriented and growth-critical discourse on what were called the 'problems of modern society' was launched. At the centre of the debate were some high-level OECD bureaucrats with strong ties to the Club of Rome, such as the Secretaries-General Thorkil Kristensen and Emile van Lennep, and the Director of Scientific Affairs Alexander King. Driven by the events of 1968 and the seemingly interrelated phenomena of social, political and environmental crises and the negative by-products of technology and industrialisation, they questioned the potential of existing political institutions to catalyse a global debate on the detrimental social and ecological effects of uncontrolled growth, the spread of technology, and individual and social alienation, which they deemed necessary. They regarded many of the existing institutions as 'guardians of the status quo and hence the enemies of change', simply because they had only 'post-facto mechanisms' available, a statement that underlines the perceived necessity of interdisciplinary systems perspectives and futures research as political technology. They were not only sceptical about the readiness of political institutions but also criticised disciplinary attitudes, 'the extraordinary arrogance of the economist, the naïvety of the natural scientist, the ignorance of the politician, and the bloody-mindedness of the bureaucrat', all unable to tackle the ensemble of problems they had identified (Alexander King 1970, c.f. Schmelzer 2017, 248). While the Club of Rome would choose global modelling as a tool for public intervention and discussion of the problematique, the OECD set up a directorate for environmental policies, at a time when no member country had an environmental ministry, and started an ambitious programme to develop alternative indicators intended to measure progress towards increasing the 'quality of life' (Schmelzer 2017, 308). Yet, while the latter failed and the OECD would – during the backlash of the 1970s – launch environmental policies that were ultimately compatible with the growth paradigm, these heterodox debates within the OECD provide an instructive case about the possibilities of autonomous supranational bureaucracies and their ability to form a platform for critical intellectual and political intervention through problematisations that challenge existing orders and paradigms (Cox 1981). Historical research can reveal the unexpected contexts of some of these proposed alternative and heterodox views and contribute an analysis of how critical interventions have been made and the reasons for if they could, or could not, prevail.

Another initiative that deserves to be mentioned here as an example of a specifically new 'détente mode' of scientific internationalism was the establishment of the International Institute for Applied Systems Analysis (IIASA). The non-governmental, international think tank opened its doors in 1973 in Laxenburg, Vienna and was funded initially by 12 national member organisations (mostly the science academies) from both sides of the Iron Curtain.[15] The initiative dated back to the mid 1960s, when President Lyndon B. Johnson launched the policies of *bridge-building* towards the Soviet Union and Eastern Europe, designed to resolve international tensions and to develop transnational relations throughout the (post-)industrialised world. Among the initiatives pursued was the establishment of a joint research centre as a site for practical, scientific collaboration between scientists and policy experts with the goal of developing mutually compatible policy expertise. As mentioned above, transnational scientific and intellectual networks have been characteristic features in the intellectual landscape at a time when various protagonists shared deeper concerns about the future development of modern, industrialised societies in rapidly changing techno-scientific environments, interdependent economies and shifting values. The establishment of IIASA is perhaps one of the most obvious examples of these emergent discussions and the development of international approaches to the co-production of knowledge and expertise. The initiative was launched at a press conference at the White House in December 1966, when the idea of a joint institute with the Soviet Union and other 'industrialised nations in East and West Europe and elsewhere' was presented to a wider public by the president of the Ford Foundation, McGeorge Bundy, who was commissioned by President Johnson

---

15 These 12 included the USA, the Soviet Union, the UK, France, Italy, Poland, Czechoslovakia, Bulgaria, West Germany, East Germany, Canada and Japan. During the 1970s five more countries were to join: Austria in 1973, Hungary in 1974, and Sweden, Finland and the Netherlands in 1976. On the history of IIASA see Riska-Campbell 2011 and Rindzevičiūtė 2016; on détente science see Graf 2017.

to pursue the project. He told the press: 'The kind of problem we are dealing with here is that all advanced economies share the problems of efficiently managing large and complicated enterprises [...] We do think that [...] if we could quietly make progress in this kind of exchange of knowledge and knowhow and have exploration in these fields of peaceful activity in advanced societies, it would be a contribution [...] to the wider cause of international understanding and of peace. [...] The problem is to take this clear fact of great common concern in matters that have to do with the business of living in an advanced society, or an advancing society, and see whether there aren't ways of setting up a new framework or a new institution or a new focus in which more progress can be made for the benefit of all.'[16] What we can understand from the wording is that the joint research centre was to be designed as an institutional response to the emerging new worldview of multifaceted power and interdependency. Collaborative research would help to develop common understandings through the study of problems arising from increasingly interdependent societies in a globalising environment. And while some of the protagonists certainly envisioned IIASA to become some kind of Cold War tool that would allow the exertion of a rather unilateral influence and transfer of systems expertise and management knowledge in order to maintain the prerogative of interpretation – or even to erode and dissolve communist ideology – the actual coproduction of 'common problems of advanced societies' unfolding at IIASA through collaborative, interdisciplinary work allowed scientific internationalism to evolve into something new.

It took six years for IIASA to open its doors, during which the initiative had been developed further in politically delicate negotiations. The notion of the 'common problems of advanced societies' provided a sustainable if abstract rationale, but the focus on problems and applied research would later also structure the institute's research matrix. Research projects at IIASA conducted by multidisciplinary teams of social and natural scientists, systems analysts, mathematicians and policy experts from both sides of the Iron Curtain would focus on complex long-term problems that similarly occurred in all advanced societies, such as problems of transboundary environmental pollution, sustainable energy supply, and urban and infrastructure plan-

---

16 Bundy, McGeorge, George Christian, and Francis M. Bator. 'News Conference at the White House 673-A', December 15, 1966. NSF Subject File: East-West Institute. Lyndon B. Johnson Presidential Library, Austin, TX; Johnson 1966.

ning, but also on the interdisciplinary co-production of frameworks to shape knowledge and expertise on newly emerging global issues such as climate change and world population. The institute became a central spot where the systems approach was developed further into an innovative and increasingly reflexive, performative and post-positivist policy science. It was combined with more profound research from various disciplines as a large number of internationally renowned scientists – ecologists, physicists, economists, sociologists, etc. – worked at IIASA. At the institute systems analysis met a complex institutional structure and the objectives were usually not defined by a single client. The projects often had multilateral and even global dimensions and the conflict potential was obvious. The actual conflicts, however, often didn't manifest themselves along the ideological lines between East and West, but rather between different disciplinary perspectives, epistemologies, attitudes towards the future and towards the properties and status of systems. On the project level such inconsistency could be turned into an asset. IIASA researchers developed, for example, methodologies such as multi-objective decision-making, participatory methods, integrated modelling approaches and comparative case studies. In contrast to previous systems approaches they attempted to take the social, political and institutional aspects of systems analysis more seriously into account, for instance the plural rationalities of the various stakeholders affected by policy-making and transformation processes or the importance of previously neglected issues such as risk and uncertainty (Duller 2016). IIASA's lasting impact and legacy lies in the provision of a sometimes contested but often innovative environment for the collaborative coproduction of common problems. A depoliticised systems approach allowed for international collaboration, mutual learning and varieties of boundary transgressions, in which disciplinary perspectives, trainings and subjectivities were made explicit and sometimes put aside in order to generate novel responses to the challenges of late modern societies (Rindzevičiūtė 2016). As a result of these collaborations numerous novel interdisciplinary and multilateral perspectives emerged at IIASA – among other places – that broadened the scope of questions to be dealt with on a scientific basis: especially, as there was often no exchange or joint problematisation at a political level on issues such as transboundary pollution, the challenges of technological change and associated risks and the problem of sustainable development.

Conclusion

Egle Rindzevičiūtė has recently argued that the mobilisation of complex systems perspectives and the 'smuggling' of policy sciences developed at IIASA, entailing notions of uncertainty and limits to knowing, had a liberalising impact on Soviet governance in the sense that they challenged totalitarian notions of control (Rindzevičiūtė 2016: 206-209). In a similar vein we can describe how these joint research initiatives opened up an international space for the construction and governance of transboundary problems in Europe, or how particular problematisations and the enactment of more ambiguous epistemologies and proposed frameworks to think about sustainable future pathways of modern societies both confirmed and challenged values, epistemic ideologies and imaginations of modernity (Schrickel 2017). In that sense we can evaluate problem-oriented research and interdisciplinary thinking at IIASA and other places in terms of the questions and futures perspectives generated and the interventions and differences they made (Barry/Born 2013). In any case, they represent various ways of actively creating and engaging with the future outlooks of modern societies in the light of emerging global challenges.

This paper presented an integrative historical approach to the study of changing conceptual frameworks and epistemological developments in interdisciplinary scientific fields such as systems science, futures research and policy sciences, which have been traced through the making of novel institutions in the years around 1970. It attempted to embed conceptual developments in the scholarly world in a broader intellectual and political environment fostering increasingly reflexive and constructive approaches. The positivist approach to progress, growth and development, which had been carefully constructed in the 19th and early 20th centuries, was called into question. The same applies to teleological ideas about the future and societal progress. It has been suggested to study these epistemic shifts through the changing notion of the *problem*, which was brought forward in various contexts at the time as a post-positivist operational concept. The conjuncture of and 'problem-talk' has been discussed as a marker for a questioning situation in the international history of modern societies, and it has been argued that through the construction of problems broader intellectual debates on the perceived challenges for modern societies, in the light of social, political and technological developments, have been enacted. In a historical moment of post-Cold War uncertainty, complexity and openness that came to the fore in the writings and discussions of various scholars, institutional entrepreneurs and policy advisors, the high-modernist problem-solving mentality of the previous era had lost its epistemic appeal.

References

Agar, Jon (2008): 'What Happened in the Sixties?' In: The British Journal for the History of Science 41/4, pp. 567-600. https://doi.org/10.1017/S0007087408001179

Allison, Graham T. (1969): 'Conceptual Models and the Cuban Missile Crisis.' In: The American Political Science Review 63/3, pp. 689-718. https://doi.org/10.1017/S000305540025853X, https://doi.org/10.2307/1954423

Anderson, Stanford (ed.) (1968): Planning for Diversity and Choice: Possible Futures and their Relations to the Man-Controlled Environment, Cambridge, MA: MIT Press.

Andersson, Jenny/Rindzevičiūtė, Eglė (2015): The Struggle for the Long-Term in Transnational Science and Politics: Forging the Future, London: Routledge. https://doi.org/10.4324/9781315717920

Andersson, Jenny (2018): The Future of the World: Futurology, Futurists, and the Struggle for the Post Cold War Imagination, Oxford: Oxford University Press.

Apostel, Léo (1972): Interdisciplinarity: Problems of Teaching and Research in Universities, Paris: Organisation for Economic Co-operation and Development.

Aron, Raymond (1963): Dix Huit Leçons sur la Société Industrielle, Paris: Idées/Gallimard.

Aronova, Elena A. (2012a): Studies of Science before 'Science Studies': Cold War and the Politics of Science in the US, UK, and USSR, 1950s-1970s, Dissertation, UC San Diego.

Aronova, Elena (2012b): 'The Congress for Cultural Freedom, Minerva, and the Quest for Instituting 'Science Studies' in the Age of Cold War.' In: Minerva 50/3, pp. 307-337. https://doi.org/10.1007/s11024-012-9206-6

Barry, Andrew/Born, Georgina (eds.) (2013): Interdisciplinarity: Reconfigurations of the Social and Natural Sciences, London: Routledge. https://doi.org/10.4324/9780203584279

Bell, Daniel (1973): The Coming of Post-Industrial Society, New York: Basic Books.
Boguslaw, Robert (1965): The New Utopians, a Study of System Design and Social Change. Englewood Cliffs, N.J.: Prentice-Hall.
Boulding, Kenneth E. (1966): 'The Economics of the Coming Spaceship Earth.' In: Environmental Quality in a Growing Economy: Essays from the Sixth RFF Forum, Baltimore: Johns Hopkins University Press.
Bowden, Sean (2018): 'An Anti-Positivist Conception of Problems.' In: Angelaki 23/2, pp. 45-63. https://doi.org/10.1080/0969725X.2018.1451461
Brick, Howard (2000): Age of Contradiction: American Thought and Culture in the 1960s, Ithaca: Cornell University Press.
Brzezinski, Zbigniew (1970): Between Two Ages: America's Role in the Technetronic Era, New York: Viking Press.
Brzezinski, Zbigniew/Huntington, Samuel P. (1964): Political Power: USA/USSR, New York: Viking Press.
Buchholz, Arnold (1968): Die große Transformation. Stuttgart: Deutsche Verlagsanstalt.
Bundy, McGeorge/George Christian/Francis M. Bator (1966): 'News Conference at the White House 673-A,' December 15, 1966. NSF Subject File: East-West Institute. Lyndon B. Johnson Presidential Library, Austin, TX.
Christiansen, Samantha/Scarlett, Zachary A. (2013): The Third World in the Global 1960s, Oxford: Berghahn Books.
Churchman, C. West (1967): 'Guest Editorial: Wicked Problems.' In: Management Science 14/ 4, B141-42.
Churchman, C. West (1978): Survey of the Contributions of the International Institute for Applied Systems Analysis to Methods other than Applied Mathematics, IIASA Archive Laxenburg.
Cosgrove, Denis (2001): Apollo's Eye: A Cartographic Genealogy of the Earth in the Western Imagination, Baltimore: Johns Hopkins University Press.
Cox, Robert W./Jacobson, Harold K. (1973): The Anatomy of Influence: Decision Making in International Organizations, New Haven: Yale University Press.
Cox, Robert W. (1981): 'Social Forces, States and World Orders: Beyond International Relations Theory.' In: Millennium – Journal of International Studies 10/2, pp. 126-55. https://doi.org/10.1177/03058298810100020501
Daly, Herman E. (1974): 'The Economics of the Steady State.' In: American Economic Review 64/2, pp. 15-21.

Deleuze, Gilles (1968): Différence et Répétition, Paris: Presses Universitaires de France.
Dery, David (1984): Problem Definition in Policy Analysis, Lawrence: University Press of Kansas.
Doel, Ronald E./Wang, Zuoyue (2001): 'Science and Technology in American Foreign Policy.' In: Alexander DeConde/Richard Dean Burns/Fredrik Logevall (Eds.): Encyclopedia of American Foreign Policy, New York: Charles Scribner's Sons, pp. 443-59.
Drucker, Peter F. (1950): The New Society: The Anatomy of Industrial Order, New York: Harper & Brothers.
Duller, Matthias (2016): 'Internationalization of Cold War Systems Analysis: RAND, IIASA and the Institutional Reasons for Methodological Change.' In: History of the Human Sciences 29/4-5, pp. 172-190. https://doi.org/10.1177/0952695116667882
During, Elie (2004): '"A History of Problems": Bergson and the French Epistemological Tradition.' In: Journal of the British Society for Phenomenology 35/1, pp. 4-23. https://doi.org/10.1080/00071773.2004.11007419
Edwards, Paul N. (1996): Closed World: Computers and the Politics of Discourse in Cold War America, Cambridge, MA: MIT Press. https://doi.org/10.7551/mitpress/1871.001.0001
Edwards, Paul N. (2010): A Vast Machine: Computer Models, Climate Data, and the Politics of Global Warming, Cambridge, MA: MIT Press.
Ellul, Jacques (1954): La technique, ou, L'enjeu du siècle, Paris: A. Colin.
Engerman, David C. (2010): 'Social Science in the Cold War.' In: Isis 101/2, pp. 393-400. https://doi.org/10.1086/653106
Erdur, Onur (2018): Die epistemologischen Jahre: Philosophie und Biologie in Frankreich, 1960-1980, Zürich: Chronos.
Esposito, Fernando (ed.) (2017): Zeitenwandel: Transformationen geschichtlicher Zeitlichkeit nach dem Boom, Göttingen: Vandenhoeck & Ruprecht. https://doi.org/10.13109/9783666301001
Ferguson, Niall/Maier, Charles S./Manela, Erez/Sargent, Daniel J. (eds.) (2011): The Shock of the Global: The 1970s in Perspective, Cambridge, MA: Belknap Press. https://doi.org/10.2307/j.ctvrs8zfp
Freire, Paulo (1970): Pedagogy of the Oppressed, New York: Herder and Herder.
Fuller, Steve (2000): Thomas Kuhn: A Philosophical History for Our Times, Chicago: University of Chicago Press.

Galbraith, John Kenneth (1968): The New Industrial State, London: Penguin Books.
Galtung, Johan (1970): Images of the World in the Year 2000, Wien: European Coordination Centre for Research and Documentation in Social Science.
Georgescu-Roegen, Nicholas (1971): The Entropy Law and the Economic Process, Cambridge, MA: Harvard University Press.
Gestwa, Klaus/Stefan Rohdewald (2009): Kooperation trotz Konfrontation – Wissenschaft und Technik im Kalten Krieg (Special Issue). In: Osteuropa 10/59.
Gilman, Nils (2003): Mandarins of the Future: Modernization Theory in Cold War America, Baltimore: Johns Hopkins University Press.
Gilman, Nils (2016): 'The Cold War as Intellectual Force Field.' In: Modern Intellectual History 13/2, pp. 507-23. https://doi.org/10.1017/S1479 244314000420
Goffman, Erving (1974): Frame Analysis: An Essay on the Organization of Experience, New York: Harper & Row.
Graf, Rüdiger (2017): 'Détente Science? Transformations of Knowledge and Expertise in the 1970s.' In: Centaurus 59/1-2, pp. 10-25. https://doi.org/10.1111/1600-0498.12148
Habermas, Jürgen (1962): Strukturwandel der Öffentlichkeit: Untersuchungen zu einer Kategorie der bürgerlichen Gesellschaft, Neuwied: Luchterhand.
Hardin, Garrett (1968): 'The Tragedy of the Commons.' In: Science 162/3859, pp. 1243-8. https://doi.org/10.1126/science.162.3859.1243
Helmer-Hirschberg, Olaf (1966): Social Technology. New York: Basic Books.
Heyck, Hunter (2015): Age of System: Understanding the Development of Modern Social Science, Baltimore: Johns Hopkins University Press.
Heymann, Matthias (2009): 'Klimakonstruktionen.' In: NTM Zeitschrift für Geschichte der Wissenschaften, Technik und Medizin 17/2, pp. 171-97. https://doi.org/10.1007/s00048-009-0336-3
Heymann, Matthias (2017): '1970s: Turn of an Era in the History of Science?' In: Centaurus 59/1-2, pp. 1-9. https://doi.org/10.1111/1600-0498.12146
Hobsbawm, Eric J. (1994): Age of Extremes: The Short Twentieth Century, 1914-1991, London: Michael Joseph.
Hoos, Ida Russakoff (1972): Systems Analysis in Public Policy: A Critique. Berkeley, CA: University of California Press.

Holling, Crawford S. (ed.) (1978): Adaptive Environmental Assessment and Management. Chichester: John Wiley & Sons.

Hughes, Agatha C./Hughes Thomas P. (2000): Systems, Experts, and Computers: The Systems Approach in Management and Engineering, World War II and After, Cambridge, MA: MIT Press.

Huntington, Samuel P. (1971): 'The Change to Change: Modernization, Development, and Politics.' In: Comparative Politics 3/3, pp. 283-322. https://doi.org/10.2307/421470

Inglehart, Ronald (1977): The Silent Revolution, Princeton: Princeton University Press.

Isaac, Joel (2012a): Working Knowledge: Making the Human Sciences from Parsons to Kuhn, Cambridge, MA: Harvard University Press. https://doi.org/10.4159/harvard.9780674065222

Isaac, Joel (2012b): 'Epistemic Design: Theory and Data in Harvard's Department of Social Relation.' In: Mark Solovey/Hamilton Cravens (eds.), Cold War Social Science: Knowledge Production, Liberal Democracy, and Human Nature, New York: Palgrave Macmillan, pp. 79-95.

Jarausch, Konrad H. (ed.) (2008): Das Ende der Zuversicht? Die siebziger Jahre als Geschichte. Göttingen: Vandenhoeck & Ruprecht. https://doi.org/10.13109/9783666361531

Jardini, David R. (1996): Out of the Blue Yonder: The Rand Corporation's Diversification into Social Welfare Research, 1946-1968, Dissertation, Carnegie Mellon University.

Jasanoff, Sheila (2001): 'Image and Imagination: The Formation of Global Environmental Consciousness.' In: Paul N. Edwards/Clark A. Miller (eds.), Changing the Atmosphere: Expert Knowledge and Environmental Governance, Cambridge, MA: MIT Press, pp. 309-337.

Judt, Tony (2005): Postwar: A History of Europe Since 1945, London: Penguin Books.

Jungk, Robert/Müllert, Norbert R. (1987): Future Workshops: How to Create Desirabl Scientific Revolutions, Chicago: University of Chicago Press.

Leendertz, Ariane (2013): '"Finalisierung der Wissenschaft"': Wissenschaftstheorie in den politischen Deutungskämpfen der Bonner Republik.' In: Mittelweg 36 22/4, pp. 93-121.

Leendertz, Ariane/Meteling, Wencke (eds.) (2016): Die Neue Wirklichkeit: Semantische Neuvermessungen und Politik seit den 1970er-Jahren, Frankfurt: Campus.

Leendertz, Ariane (2019): 'Amerikanische Policy-Forschung, Komplexität und die Krise des Regierens: Zur gesellschaftlichen Einbettung sozialwissenschaftlicher Begriffsbildung.' In: Berichte zur Wissenschaftsgeschichte 42/1, pp. 43-63. https://doi.org/10.1002/bewi.201901879

Leibetseder, Bettina (2011): 'A Critical Review on the Concept of Social Technology.' In: Social Technologies 1/1, pp. 7-24.

Light, Jennifer S. (2004): From Warfare to Welfare: Defense Intellectuals and Urban Problems in Cold War America, Baltimore: Johns Hopkins University Press.

Lorenz, Chris/Bevernage, Berber (eds.) (2013): Breaking Up Time: Negotiating the Borders Between Present, Past and Future, Göttingen: Vandenhoeck & Ruprecht. https://doi.org/10.13109/9783666310461

Luhmann, Niklas (1972): 'Weltzeit und Systemgeschichte.' In: Peter Christian Ludz (ed.), Soziologie und Sozialgeschichte: Aspekte und Probleme, Wiesbaden: VS Verlag für Sozialwissenschaften, pp. 81-115. https://doi.org/10.1007/978-3-322-83551-2_4

McCombs, Maxwell E./Shaw, Donald L. (1972): 'The Agenda-Setting Function of Mass Media.' In: Public Opinion Quarterly 36/2, pp. 176-187. https://doi.org/10.1086/267990

McLuhan, Marshall (1964): Understanding Media: The Extensions of Man, New York: McGraw-Hill.

Meadows, Donella H. et al. (1972): The Limits to Growth: A Report for the Club of Rome's Project on the Predicament of Mankind, New York: Universe Books. https://doi.org/10.1349/ddlp.1

Mittelstraß, Jürgen et al. (eds.) (2005-2016): Enzyklopädie Philosophie und Wissenschaftstheorie, 6 Volumes, Stuttgart: Metzler. https://doi.org/10.1007/978-3-476-00134-4

Nelkin, Dorothy (1979): 'The Social Responsibility of Scientists.' In: Annals of the New York Academy of Sciences 334/1, pp. 176-82. https://doi.org/10.1111/j.1749-6632.1979.tb53673.x

Novick, Peter (1988): That Noble Dream: The 'Objectivity Question' and the American Historical Profession, Cambridge, UK: Cambridge University Press. https://doi.org/10.1017/CBO9780511816345

Osborne, Peter (2015): 'Problematizing Disciplinarity, Transdisciplinary Problematics.' In: Theory, Culture & Society 32/5-6, pp. 3-35. https://doi.org/10.1177/0263276415592245

Özbekhan, Hasan (1969): ‚Toward a General Theory of Planning.' In: Erich Jantsch (ed.), Perspectives of Planning: Proceedings of the OECD Working Symposium on Long-Range Forecasting and Planning, Bellagio, Italy 27 Oct – 2 Nov 1968, Paris: OECD, pp. 47-159.

Özbekhan, Hasan (1970): 'The Predicament of Mankind.' In: C. West Churchman/Richard O. Mason (eds.), World Modeling: A Dialogue, Amsterdam: North-Holland, pp. 11-25.

Peccei, Aurelio (1969): The Chasm Ahead, London: Collier Macmillan Ltd.

Philp, Mark (2007): Political Conduct, Cambridge, MA: Harvard University Press.

Radkau, Joachim (2014): The Age of Ecology, Cambridge, UK: Polity.

Reisch, George A. (2016): 'Telegrams and Paradigms: On Cold War Geopolitics and The Structure of Scientific Revolutions.' In: Elena Aronova/Simone Turchetti (eds.), Science Studies During the Cold War and Beyond – Paradigms Defected, New York: Palgrave Macmillan, pp. 23-53. https://doi.org/10.1057/978-1-137-55943-2_2

Rheinberger, Hans-Jörg (2010): On Historicizing Epistemology: An Essay, Stanford, CA: Stanford University Press.

Richta, Radovan (1969): Civilization at the Crossroads; Social and Human Implications of the Scientific and Technological Revolution, Abingdon: Routledge.

Riska-Campbell, Leena (2011): Bridging East and West: The Establishment of the International Institute for Applied Systems Analysis (IIASA) in the United States Foreign Policy of Bridge Building, 1964-1972, Helsinki: The Finnish Society of Science and Letters.

Rindzevičiūtė, Eglė (2015): 'The Future as an Intellectual Technology in the Soviet Union: From Centralised Planning to Reflexive Management.' In: Cahiers Du Monde Russe 56/1, pp. 111-34. https://doi.org/10.4000/monderusse.8169

Rindzevičiūtė, Eglė (2016): The Power of Systems: How Policy Sciences Opened Up the Cold War World, Ithaca: Cornell University Press. https://doi.org/10.7591/cornell/9781501703188.001.0001

Rittel, Horst W. J./Webber, Melvin M. (1973): 'Dilemmas in a General Theory of Planning.' Policy Sciences 4/2, pp. 155-69. https://doi.org/10.1007/BF01405730

Rodgers, Daniel T. (2011): Age of Fracture, Cambridge, MA: Harvard University Press.

Rohde, Joy (2013): Armed with Expertise: The Militarization of American Social Research during the Cold War, Ithaca: Cornell University Press. https://doi.org/10.7591/cornell/9780801449673.001.0001

Rohde, Joy (2017): 'Pax Technologica: Computers, International Affairs, and Human Reason in the Cold War.' In: Isis 108/4, pp. 792-813. https://doi.org/10.1086/695679

Rosenau, James N. (ed.) (1969): Linkage Politics: Essays on the Convergence of National and International Systems. New York: Free Press.

Schaffer, Simon (2013): 'How Disciplines Look.' In: Andrew Barry/Georgina Born (eds.), Interdisciplinarity. Reconfigurations of the Social and Natural Sciences, London: Routledge, pp. 57-81.

Schmelzer, Matthias (2012): 'The Crisis before the Crisis: The "Problems of Modern Society" and the OECD, 1968-74.' In: European Review of History: Revue Européenne d'histoire 19/6, pp. 999-1020. https://doi.org/10.1080/13507486.2012.739148

Schmelzer, Matthias (2017): '"Born in the Corridors of the OECD": The Forgotten Origins of the Club of Rome, Transnational Networks, and the 1970s in Global History.' In: Journal of Global History 12/1, pp. 26-48. https://doi.org/10.1017/S1740022816000322

Schrickel, Isabell (2017): 'Control versus Complexity: Approaches to the Carbon Dioxide Problem at IIASA.' In: Berichte zur Wissenschaftsgeschichte 40/2, pp. 140-159. https://doi.org/10.1002/bewi.201701821

Schumacher, Ernst F. (1973): Small Is Beautiful: Economics as if People Mattered, New York: Blond & Briggs.

Seefried, Elke (2015a): Zukünfte: Aufstieg und Krise der Zukunftsforschung 1945-1980, Berlin: Walter de Gruyter. https://doi.org/10.1515/9783110349122

Seefried, Elke (2015b): 'Rethinking Progress. On the Origin of the Modern Sustainability Discourse, 1970-2000.' In: Journal of Modern European History 13/3, pp. 377-400. https://doi.org/10.17104/1611-8944-2015-3-377

Shapin, Steven (2009): The Scientific Life: A Moral History of a Late Modern Vocation, Chicago: University of Chicago Press. https://doi.org/10.7208/chicago/9780226750170.001.0001

Solovey, Mark (2001): 'Project Camelot and the 1960s Epistemological Revolution: Rethinking the Politics-Patronage-Social Science Nexus.' In: Social Studies of Science 31/2 (2001), pp. 171-206. https://doi.org/10.1177/030631270031002003

Sommer, Vítzslav (2017): "'Are We Still Behaving as Revolutionaries?'": Radovan Richta, Theory of Revolution and Dilemmas of Reform Communism in Czechoslovakia.' In: Studies in East European Thought 69/1, pp. 93-110. https://doi.org/10.1007/s11212-017-9280-2

Sorokin, Pitirim A. (1960): 'Soziologische und Kulturelle Annäherungen zwischen den Vereinigten Staaten und der Sowjetunion.' In: Zeitschrift für Politik: Organ der Hochschule für Politik München 7/4, pp. 341-370.

Sprenger, Florian (2019): Epistemologien des Umgebens: Zur Geschichte, Ökologie und Biopolitik künstlicher environments, Bielefeld: transcript.

Suri, Jeremi (2009): Power and Protest: Global Revolution and the Rise of Detente, Cambridge, MA: Harvard University Press.

Taylor, Mark C. (2001): The Moment of Complexity: Emerging Network Culture, Chicago: University of Chicago Press.

Thomas, William (2015): Rational Action: The Sciences of Policy in Britain and America, 1940-1960, Cambridge, MA: MIT Press. https://doi.org/10.7551/mitpress/9997.001.0001

Tichenor, Phillip. J et al. (1970): 'Mass Media Flow and Differential Growth in Knowledge.' In: Public Opinion Quarterly 34/2, pp. 159-170. https://doi.org/10.1086/267786

Tinbergen, Jan (1961): 'Do Communist and Free Economies Show a Converging Pattern?' In: Soviet Studies 12/4, pp. 333-341. https://doi.org/10.1080/09668136108410255

Touraine, Alain (1971): The Post-Industrial Society. Tomorrow's Social History: Classes, Conflicts and Culture in the Programmed Society. New York: Random House.

Vieille Blanchard, Elodie (2010): 'Modelling the Future: An Overview of the "Limits to Growth" Debate.' In: Centaurus 52/2, pp. 91-116. https://doi.org/10.1111/j.1600-0498.2010.00173.x

Warde, Paul/Robin, Libby/Sörlin, Sverker (2019): The Environment: A History of the Idea, Baltimore: Johns Hopkins University Press.

White, H. Theodore (1967): 'The Action Intellectuals.' In: Life Magazine, June 9.

Wirsching, Andreas/Therborn, Göran/Eley, Geoff/Kaelble, Hartmut/Chassaigne, Philippe (2011): 'The 1970s and 1980s as a Turning Point in European History?' In: Journal of Modern European History 9/1, pp. 8-26. https://doi.org/10.17104/1611-8944_2011_1_8

The Problematic of Transdisciplinary Sustainability Sciences

Esther Meyer

Introduction

Sustainable Development (SD) finds its discursive breakthrough in 1987 through the final report of the Brundtland Commission, *Our Common Future* (Vanhulst/Beling 2014; Sneddorn et al. 2006). The Brundtland report substantially conveys the regulative specification of a worldwide social and ecological national economic development, justified by the possibility of equal opportunities also for future generations (intergenerational justice). In addition, this development should be designed in such a way that equal access to resources for all living people is possible (intragenerational justice) (Hauff 1987; Dingler 2003). Reactions to the report reveal the nature of its global regulatory appeal, because intra- and inter-generational justice can only be defined according to political values (Vanhulst/Beling 2014; Grunwald 2011). In 2015, the United Nations set the *Sustainable Development Goals* (SDGs), differentiating and equating SD explicitly with peace and security, natural and cultural diversity, democracy, eradicating poverty, as well as equal rights and opportunities for women and men (SDGs 2015). SD simultaneously tends to be shaped by a *hegemonic discourse of SD* (Hajer 1995; Höhler/Luks 2004; Brown 2016; Vanhulst/Zaccai 2016; Albán/Rosero 2016)[1] that ultimately

---

[1] Discourse understandings, or the different discursive analytical orientations of the authors who refer to the hegemonic discourse of SD, are not discussed here. My own, previously carried out, discourse-analytical research (Meyer 2020) is based on the understanding of critical discourse analysis. According to Adele Clarke (2012) critical discourse analysis pays special attention to the ways in which dominant theories emerge and, through their discourses, (re)produce power relations.

counteracts SD as it is envisaged by the SDGs. Around the 1990s, so-called *sustainability sciences* began to form and characterise themselves as inter- or transdisciplinary. Sustainability sciences are constituted by and respond to international sustainability politics and organisations, intertwined with hegemonic political interests. Transdisciplinary (td) sustainability sciences especially aim to generate topics and research questions in collaboration with representatives of diverse societal groups in order to identify pressing sustainability problems. Accordingly, questions arise concerning the entanglement with and positioning towards a superordinate hegemonic discourse of SD. Thus, transformative and interventionist approaches to exploring a sustainable cohabitation are being hampered. The questions arise, firstly, if, and, secondly, which theories towards societal transformation are missing in td sustainability sciences, and how may sustainability and td sustainability research be re-invented in order to explore and shape a sustainable cohabitation.

My contribution starts with my methodology, the problematisation of 'notions of problems' (Bowden/Kelly 2018: 3). After the introduction of the methodology follows an outline of the hegemonic discourse of SD and the consequences it produces. That leads to the introduction to td sustainability sciences. The objective is to analyze how problematisations in td sustainability sciences relate to concepts that have emerged through the hegemonic discourse of sustainability. In sustainability sciences, I suggest this is the concept of *challenge*. While the first part deals with the problematic *of* (td) sustainability sciences, the second part deals with the problematic *in* td sustainability research. The differentiated addressing of the problematic deals with methodological considerations and experiments for a td sustainability research that is aware of its entanglement of epistemological and normative dimensions. The aim of my research is to explicate reproducing discourses and constructions of handling problems in td sustainability sciences that suppress the subversive potential of radical transdisciplinary logics and comprehensions of a *generative problematic in td sustainability research*.

Methodology: problematisation of problem understandings

Transdisciplinarity and td sustainability research can gain significance as a counter project to the *hegemonic discourse of SD*. However, td sustainability sciences are partly interwoven with the hegemonic discourse. Being a relatively new phenomenon within the discourse, their efficacy is limited from the outset by existing power relations. It is here that the problematic unfolds itself as a possibility to work with. The problematic of td sustainability research can be found in its in between position amid distinct, inconsistent, contradictory paradigms. Td sustainability sciences are, as Michel Foucault would say, 'always inscribed in a game of power, but always also a limitation or rather: bound to the limits of knowledge, which emerge from it, but nevertheless condition it' (Foucault 1978: 123, in Bührmann/Schneider 2008: 53, my translation).

The concept of *problem* has a major bearing on td sustainability sciences. On the one hand, td sustainability sciences tend to be considered as ethical and intellectual revolutions or innovations in the mode of thought and, thus, as solutions to sustaining global social-ecological problems. On the other hand, these problems persist and accumulate due to another hegemonic economic-political level that is often overlooked in research practice. These problems then tend to be at the same time the condition of possibility for td sustainability sciences to be constituted, legitimised, and made possible. The meanings of problems and their function for td sustainability sciences therefore seem to constitute their problematic. Starting from a *problematic constitution of problems* 'offers heuristic notions that allow the reformulation of the manner in which problems are conceived' and, as Maria Kaika further writes concerning a radical political ecology, '[t]his inclusive approach does not place itself on "managerial" ground' (Kaika 2003, in Blanchon/Graefe 2012: 47), but on a philosophical movement to pose different research questions and other problems to be investigated (Bachelard 2012; Maniglier 2019). In which contexts of meaning are problems posed? What would be a different theorisation of the problem? With Foucault problematisation means to carve out conditions of possibilities that enable different solutions to symptomatic problems (Defert/Ewald 2005). By scrutinising supposed solutions in td sustainability sciences, I will first make the problem approachable.

The problematic *of* sustainability sciences

Hegemonic discourse of sustainable development

The hegemonic discourse of SD is aligned to neoliberal forms and goals of organising (environmental) policies towards profit maximisation of market enterprises (Castree 2002). A critical discourse analysis published in 2014 by Carol Kambites examines discourse strands of SD in the respective strategy papers of British governments in the 1990s and 2000s and comes to the conclusion: 'sustainable development is presented from within the paradigm of neoliberalism and neoclassical economics' (Kambites 2014: 345). In Germany the analysis by Johannes Dingler on SD shows that a 'decrease in the stress of intragenerational justice' (Dingler 2003: 255, my translation) can be seen. 'Intragenerational justice is, thus, increasingly reduced to equality of opportunity and subsumed under market-based instruments' (ibid, my translation). At the same time intergenerational justice is prioritised, which matches well with the normative goal of having the chance of private asset protection and its intergenerational transfer. These patterns of significations of SD neglect the discourses of social redistribution within one generation.

The research project 'NEDS – Nachhaltige Entwicklung zwischen Durchsatz und Symbolik' ('Sustainable development between throughput and symbolism') analyzes the Brundtland report regarding the economic construction of ecological reality. The research project identifies seven coherent theses – thereby differentiating the thesis of the unsustainability of modernity. They outline how 'economic logic, natural and technological scientific expectations and juridical, administrative regulations intertwine and have contributed significantly to a discursive version of sustainability as a management problem' (Höhler/Luks 2004, my translation). The authors see SD shifting from an understanding of nature and ecology to an understanding of mere economically manageable and controllable environments divided into scarce resources. The hegemonic economic conception of SD is reflected in the guiding principle of *weak sustainability* (Williams/Millington 2004; Ziegler/Ott 2011), which assumes only a few, isolable sustainability dimensions, as well as their interchangeability: economic, ecological or social goals should be integrated into behaviour and economic activity. In Germany, the final report of the Enquete Commission, 'Protection of Man and the Environment', proposes a subdivision into three pillars: ecological, economic, and social (Enquete Commission of the 13th German Bundestag 1997). In addition, multi-pillar models and one-pillar models have been developed ('from one dimension to eight dimensions', Tremmel 2003: 116, my translation). Also, the cultural, the institutional and the political are mentioned as important parts (Michelsen/Adomßent 2014). Moreover, in this discourse, not all authors speak of pillars, but instead, for example, of different dimensions, like Niranji Satanarachchi and Takashi Mino (2014) or the Preamble of the SDGs. The concept of *strong sustainability* (Ott/Döring 2004; Ziegler/Ott 2011), however, is not contained within the logic of the pillar-discourse: nature as an ecological basis of life is not considered substitutable. The relative approach via goals, pillars or dimensions of sustainability has different effects as to how social or ecological target dimensions are integrated into a discourse that is governed by a priori economic ratings.

What are the problems of the hegemonic discourse of sustainable development?

Human rights, which are valid for all current and future humans (Ott 2014; SDGs 2015), count as substantial minimal goals for sustainability and thus constitute the basis of normative sustainability ethics (Carnau 2011).[2] From a sustainability ethical perspective, human behaviour can therefore be assessed on the basis of whether it is *life-sustaining* (Carnau 2011; Olssen 2014). 'The hegemonic discourse of sustainable development is in the discursive tradition of [...] modernity' (Dingler 2003: 484, my translation). The social development indicated in the discourse, however, could have led to a crisis threatening the livelihoods of today's and future people's lives ('thesis of the unsustainability of modernity', ibid: 493).[3] SD strategies, as they refer to in the Brundtland report, aim at achieving economic growth that is desirable

---

2   This work is not concerned with the definition of a normative-prescriptive ethics of sustainability. Thus, the 'future', related to human rights and climate change, remains open. The work, however, is based on the premise that a normative-prescriptive ethics of sustainability is recognised.

3   The designation of an "ecological modernization" by Hajer (1995) counts as an origin in the German- and English-speaking discourse. Likewise, the criticism of Eblinghaus and Stickler from 1996 can be mentioned. Criticism of eurocentrism and the globalisation of occidental development theories, in this respect, comes from Arturo Escobar (1995) and Wolfgang Sachs (1993; 2002). Other authors grasp the thesis of the 'unsustainability of modernity'

for as many nation states as possible in order to establish both inter- and intra-generational justice. The unit in which national economic growth is measured is the quantitatively expressed gross domestic product (GDP). This means that the goal of SD is that all countries always achieve the highest possible economic parameter – sustainable growth or *green growth* (Höhler/Luks 2004; Brand/Wissen 2017; Acosta/Brand 2018). In economic theory, higher growth figures equate to more capital being available for the state to finance environmental protection or social compensation. However, in order to achieve these growth figures, nature, the environment, resources and people – life – are subordinated to economic development and consumed in life-destroying proportions (Moore 2016). This happens in an exponential way, because of the national-economic belief in higher growth numbers as a solution and in the complete governability of social-ecological problems at local and global level. Thus, national-economic theory of this kind and its politics are dysfunctional as they cannot meet the requirements of sustainability. An analysis of the SDGs shows that sustainability-relevant norms are attributed to the local and global levels, which in turn can have effects at the national-economic meso-level, 'as the normative core and the focus of action and interventions' (Schmieg et al. 2018). However, the non-sustainable norms of the meso-state level are not problematised in the UN documents (Parenti 2016). The transnational agenda of SD, emerging at the beginning of the 1970s from environmental and justice movements, has been incorporated into the neoliberal agenda, starting in the 1980s and 1990s with more and more success. Sustainability, therefore, under the roof of SD, serves to strengthen and spread neoliberal hegemony, leaving eco-political and human rights interests in marginalised positions. If sustainability was caught in a neoliberal hegemony, fractures within the latter are showing up and might change constellations (Brown 2016). This is also reflected in the SDGs, as important documents that aim to advance sustainability (Schmieg et al. 2018), and that differ from international sustainability documents of the late 1980s, 1990s and 2000s. And, as Julien Vanhulst and Adrián Beling write, 'even if conservative understandings of SD remain dominant, they continue slowly to lose ascendancy over global debates in the discursive field of SD, as the growing emergence of alternative discourses (and their coalitions)

(Dingler 2003) or the "economic construction of ecological reality" (Höhler/Luks 2004) as a dispositive (Timpf 2000).

proves' (Vanhulst/Beling 2014: 61). The very question in and beyond this contribution is how td sustainability sciences have reacted to neoliberal history and present dynamics and, thus, relate to the hegemonic discourse of SD.

(Transdisciplinary) Sustainability sciences

Sustainability sciences themselves make up parts of complex dynamic sustainability contexts within the human-nature system and behave towards them in an evaluative and reflexive way (Satanarachchi/Mino 2014). In the constitution of sustainability sciences there are two strikingly parallel developments: on the one hand projects in the theory of science, such as transdisciplinarity, and on the other hand transnational negotiations. In both cases, it is a question of moving boundaries, in collaborations between scientific and non-scientific actors (Vilsmaier 2018; Schmidt 2011), in order to pursue SD. The spectrum of discursive events that constitute sustainability sciences is wide. For the sake of systematics they can be represented on five interwoven levels: 1) political with the UN conferences[4]; 2) theory and politics of science with concepts such as transdisciplinarity (Klein et al. 2001; Osborne 2015), mode-2 (Gibbons et al. 1994; Gibbons 1999; Nowotny et al. 2001), or post-normal science (Funtowicz/Ravetz 1993: 3) publications such as from Robert Kates and William Clark et al. in *Science* in 2001 that present sustainability sciences as a programmatic scientific research field (Kates et al. 2001); 4) non-university institutes, NGOs, civil society, companies that strongly re-

---

4 'United Nations Conference on Human Environment' in 1972, 'United Nations Conference on Environment and Development' in 1992. From these conferences emerged programmes, as the United Nations Environment Program (UNEP), the final report of the World Commission on Environment and Development and the Agenda 21, the World Climate Summits, international follow-up conferences such as the World Summit on Sustainable Development in Johannesburg in 2002, or the SDGs document in 2015. There were also counter-reactions to the European and North American 'global consensual positions on ecology and development' (Vanhulst/Beling 2014: 55). The Latin American Global Model (or Bariloche Model) (Herrera et al. 1976) replied in 1976 to the MIT report 'The limits to growth' (Meadows et al. 1972), and, in 1991, the report 'Nuestra propia agenda sobre desarrollo y medio ambiente' ('Our own agenda on development and environment') of the Development and Environment Commission of Latin America and the Caribbean was published in response to the Brundtland report and in preparation for Rio 1992 (Vanhulst/Beling 2014; Vanhulst/Hevia 2016: 178). See also Meyer/Vilsmaier 2020.

acted to 5) global events that destroyed nature and called for environmental policy measures.

Joachim Spangenberg distinguishes the understandings of sustainability sciences as being between a 'more traditional disciplinary-based science for sustainability and the transdisciplinary science of sustainability' (Spangenberg 2011: 275). Td sustainability sciences fall in the category of science *of* sustainability. This emergent mode of research is aiming at the plurality of knowledges and perspectives, as well as process orientation combined with a normative orientation towards sustainability or SD. It is criticising modern institutionalised demarcations and understandings of research, such as scientific objectivity and progress (Vilsmaier et al. 2017; Vilsmaier 2018). Research in td sustainability sciences may open up a platform on which the boundaries that constitute research are shifted (Schmidt 2011). Relationships between the scientific and non-scientific emerge, for example in consideration of traditional or local everyday knowledge (Klein 2014).

According to Julie Thompson Klein's analysis of discourses on transdisciplinarity, the dominant understanding of and lived research cultures in td sustainability sciences is attributable to the 'discourse of problem solving' (Klein 2014: 70; Schmidt 2011). The discourse is represented by the Swiss-based 'Network for Transdisciplinary Research' known as 'td-net', that was founded at a congress held in Zurich in 2000. Thus it is sometimes classified as a 'Swiss or German school of TD because the approach was signaled in the late 1980s and early 1990s in Swiss and German contexts of environmental research' (ibid: 74). The results of a collocation analysis focusing on the concept of problem in English-speaking articles of the journal *GAiA* published up to and including the year 2017 confirm that td sustainability sciences appeal to problem-solving as their normative target (Meyer 2020).

Exemplary problem understandings in transdisciplinary sustainability sciences

Controversial problem contents as justification moments for sustainability sciences

The first UN conference on the human environment in Stockholm in 1972, as a reference point for sustainability sciences, showed that what are regarded as sustainability-related societal problems is contested. The countries of the Global North in particular demanded measures to limit industrial pollution, while the countries of the Global South pushed for a catch-up of prosperity and brought forward medical and educational concerns. There were therefore different ideas about this conference, which resulted in a compromise to capture everything as environment and to conceptualise human progress with the label of SD in order to dissolve the contradiction or better emphasise the compatibility between economic growth and environmental protection (Hopwood et al. 2005; Sneddorn et al. 2006; Vanhulst/Beling 2014).[5]

Challenge

The normative background against which problems are assessed is a functioning society as a prerequisite for SD. SD itself is equated with a societal challenge. The use of the concept of challenge points to the following developments: problems associated with sustainability are labeled as societal challenge(s), replacing so-called old social problems, like hunger, illness, and poverty (Rockström et al. 2009; Jerneck et al. 2011). The sustainability challenges, in their unlocalised rhetoric, refer to expected welfare losses or gains, are uncertain, speculative, and cannot be understood by social collectives from experiences (Jerneck et al. 2011). At the same time, they are communicated as alarming due to the irretrievability of unique opportunities with advancing time (Moore 2016). Within market economy thinking challenges are connoted positively as they simultaneously offer an opportunity for innovative advancement and progress for a sustainable society, if correspond-

---

[5] The comprehensive *empirical* question about how the controversial problems found their ways into td sustainability sciences or were not assessed as relevant, must first be put aside.

ing – also positively connoted – risk-oriented performance is shown. The sustainability challenges thus fit without contradictions into the discourse of the freeing of the markets from socio-ecological policy regulations.

Sustainability challenge is a relative concept that does not diagnose any spatial and temporal limits or goals in view of future uncertainties to be speculated. Therefore the term describes the discourse of SD as a dynamic shift of boundaries or relative goal within the concept of weak sustainability. This is incompatible with the discourse on strong sustainability (Ott/Döring 2004; Ziegler/Ott 2011), which in turn *identifies* planetary boundaries (Rockström et al. 2009).[6] Within these boundaries all human endeavor and striving, the mode of economic activity, has to happen. This discourse was stronger in *Limits to Growth* (Meadows et al. 1972) and in the environmental concerns at the beginning of the 1970s (Williams/Millington 2004). What is also striking is another development that goes along with the terms of the 'problem' and 'challenge': 'dilemma' is underrepresented as a concept in sustainability-related scientific publications.[7] This term means that there is no solution that would be morally acceptable to all stakeholders – we remain terminologically in the discourse strand of the td sustainability sciences – to derive a conflict-free action. The concept of 'dilemma' indicates epistemological or ethical issues, namely, how to deal with manifold and conflicting epistemologies or moral norms, or which ethical legitimacy becomes accepted and how. These questions are not central in td sustainability sciences (Krohn et al. 2017). It therefore seems promising to work on the thesis of a repression of dilemma and conflict in discourses on sustainability and SD in light of the solution of familiar social problems with market economic strategies – rebranded as sustainability challenges. One hypothesis is that the prioritisation of intergenerational instead of intragenerational research questions and the marginalised theories dealing with differences and moral conflicts in td sustainability sciences explain each other.

After the naming of the problematic of td sustainability sciences, the next part of this chapter attempts to highlight theoretical-methodological

---

6 The Rockström et al. paper, however, leaves a space for discussion open by using the term challenge.

7 No search results (August 2018) came from the terms 'moral dilemma AND sustainab*' in the Web of Science, a relevant database of scientific publications (https://login.webofknowledge.com/).

starting points, which answer to just that epistemic-ambiguous (Harrasser/ Sohldju 2016) problematic, namely being taught to think in an even, sustainable way 'that created today's turbulence [and] is unlikely to help us solve it' (Moore 2016: 1). In so doing, the figure of the problematic, as it is envisaged in French philosophy of the 20th century, is connected to td sustainability research for further development.

The problematic *in* transdisciplinary sustainability research

In td sustainability research, moral and epistemological dimensions are interwoven. Reading about the problematic in twentieth-century French philosophy[8] raises the question of an *epistemology of the problematic* that can supplement the basic normative coordinates in sustainability and sustainability research – change and adaptation – with basic questions. Such as, how does td sustainability research understand, explain and perform relationships between and through the form of research itself, concerning individuality, collectivity, subjectivity, and objectivity? In the following, I take up Gilbert Simondon, because his thinking of the problematic can enrich conceptual approaches in the process- and change-oriented td sustainability research (Engbers 2020) that orient beyond hegemonic discourses and practices of SD.[9] With his conceptualisation of dynamics and change through close studies of the modes of functioning of the living, Simondon is able to offer a 'radically transdisciplinary' (Scott 2014: 3) alternative to a mechanical concept of *development* covered in the hegemonic discourse of SD. In contrast, the problematic becoming, or individuation, as he calls the dynamics as dimensions of the living, keeps moving in permanent relation to particular, multi-layered, multi-dimensional, interior and exterior (Voss 2018: 101) environments. Individuation describes the inventive finding of a partial own in the conditional higher social dimension, by transindividual participation (Simondon 1964/2007: 31; Voss 2018: 96, 104). The psychic and the collective

---

8 The interdisciplinary research project 'Complexity or Control? Paradigms for sustainable development', in preparation for the workshop 'Thinking the Problematic', read together several philosophical texts.

9 I have worked with a few existing translations into German and English of his work as well as secondary literature.

are constituted by individuation (Simondon 1964/2007: 36).[10] Individuation, as a structural description of the dynamics and vectors of change, is neither to be understood as a sole adapting to the specific higher social dimension, nor to be understood in such a way as to be based solely on the change of the higher social dimension. Rather, individuation is explained by inventing internal structures (Voss 2018: 95), in accordance with the changed exterior structures, and, thus, inventing a new metastable, participative and symbiotic relationship state between exteriority and interiority (Simondon 1964/2007: 35). The problematic arises through resonating the exterior in the interior (Voss 2018: 94). Individuation is an ever-inventing of new problematics and always necessary dynamics of living (Simondon 1964/2007: 36). For td sustainability research the recognition of Simondon's structure of individuation would provide the ability to interweave with an awareness of environmentality, the interior, the exterior, as well as with a different awareness of temporality, such as of the previous, and the future. The political-normative of sustainability is manifested in the dynamics of change, whereby these are to be thought of as, in different strengths, mutually conditional interwoven starting points: the interior, the relations and the exterior (Harrasser/Sohldju 2016; Voss 2018: 98). The problematic is generative and sustainable, because it cannot be resolved by an optimistic detachment from material conditionalities for the living such as the externalisation of the global dimensions of our modes of production and consumption, for example.

Sustainability sciences are based on ethics, because of their explicit normative orientation towards sustainability. Which policies of change, which collective normative movements (such as those contained in a normative-prescriptive sustainability ethics or in the SDGs), can we deduce from the dynamics of life described in this way? Where do I find the normative momentum with regard to sustainability? A normative momentum that is not assessing or defining the uncertainty of a problem-transformation with regard to fixed outcomes, nor talking of sustainability problems or challenges, but of sustainability-related events that provoke social changes to challenge td sustainability research with the question: Why and how may td sustainability researchers shape these social changes? Which normative movement can be invented in concepts 'such as ecological economics, polit-

---

10 Just as little is said of an initial psyche confronting an initial collective, epistemologically an initial juxtaposition of subject and object can be used (Maniglier 2012).

ical ecology, de-growth, ecosocialism, ecofeminism, environmental justice' (Vanhulst/Beling 2014), i.e. for the preservation of life and how to work in td sustainability research?

Methodological problematic

How can we think of methodologies for td sustainability research that are coherent with *epistemologies of the problematic* (Maniglier 2019)? How to shape conditions for transdisciplinary possibilities to unfold the problematic? How can we activate an ethical practice in td sustainability research that allows for the speculative and failure and, thus, arrive at an ecology of practices that create spaces of opportunities beyond a cling to computable decision nodes (Stengers 2010)?

The problematic in td sustainability research may be addressed by situational, contextualised decision-making and responsiveness, 'local values, traits, beliefs, and arts for action' (Fals Borda 1995), entrepreneurial creativity, humor (Savransky 2018), attitude and ethics (Meckesheimer 2013), as well as an (algorithmic) learning, which recognises temporally and spatially related, multiple different sustainability contexts and continues the resulting decisions as limiting moments, instead of universal, methodical programs (Harrasser/Sohldju 2016). Methodological approaches that go in this direction are oriented along 'a questioning perspective that does not rush for direct straightforward solutions to problems, [...] an appeal to imaginative possibilities and especially subversive imagination; a hands-on approach to experimentation which is not limited to linear logico-deductive processes [...], spaces of possibilities to play and experimentally and aesthetically engage with.' (Kagan 2015: 2) In search for a 'particular methodology in transdisciplinarity' through his Deleuze reading, Patrice Maniglier calls for 'the introduction of comparative methods across the disciplines': 'To compare consists in experiencing, within one's system of categories, a variation of the very type that functions as the heading that makes the comparison possible' (Maniglier 2019).

There are diverse and recent methodical examples and experiments that can be interpreted as problematic and transdisciplinary methodology, or that have even been designed as such: design methods (Jonas 2015; Peukert/Vilsmaier 2019), generative picturing (Brandner 2020), transformative scenario planning (Freeth/Drimie 2016), case-based mutual-learning sessions

(Vilsmaier et al. 2015), mutual listening (Meckesheimer 2013), story-telling salons (Richter/Rohnstock 2016), and remembrance work (Haug 1999). With a 'thinking practice of problematic designing', Daniela Peukert and I recently attempted to open an epistemological perspective in and for td sustainability research. It is designed to methodologically capture the experience of a problematic (Meyer/Peukert 2020) and for a multi-dimensional methodology allowing Simondon's approach to be interwoven with the complexity that sustainability and td sustainability sciences demand. In addressing questions of how we can include the conditions of our research into the research itself, we can work out dimensions in and for the respective research situation. The epistemological concept of problematic designing, as a thinking practice, together with the methodological design canon, is an invitation to expand the methodological canon of td sustainability research.[11]

Epistemologies of the problematic start at the relation to uncertainties, be they the past, the other or the future (Vilsmaier et al. 2017) and regard 'the effects themselves (as) the cause of the world's development' (Aicher 1991: 186, my translation; Harrasser/Sohldju 2016; Moore 2016). The (future as) playful-speculative remains tied back to its conditions, namely (preservation of) life itself and its 'pre-individual nature' (Voss 2018: 96).

That calls for an ethical research practice, protected against neoliberal re-enclosure (Meckesheimer 2013; Strong et al. 2016) to enable td researchers to make decisions without competitive pressure and not to set numerical optimal solutions but an 'ecology of practices' as a standard (Stengers 2005; 2010). The speculative is therefore no challenge to climb the highest mountain but to invent other mountain worlds. Td sustainability research must distinguish itself from a concept of science that evaluates the progress of knowledge, as well as researchers on the basis of an impact factor (Schmidt 2011) and that always excludes other forms of research (Meckesheimer 2013), as well as unpredictable insights – which, however, are relevant to sustainability research and, thus, to sustainability. As Andreas Kläy et al. ask in the journal *Futures*: 'Science for sustainable development is, thus, confronted with a fundamental contradiction arising from this double normative framing of science policy: can scientists really live up to their role of contributing to sustainable development, while at the same time helping societies achieve

---

11 Daniela Peukert is currently working on this topic as part of her PhD, see https://www.danielapeukert.de/.

only greater economic growth, at the expense of equity and the environment?' (Kläy et al. 2015: 73)

Conclusions

The idea of sustainability allows us to reflexively refer to different ways of life on planet Earth with regard to our own behaviour and at the same time renegotiates the material conditionality for these ways of life. Being normatively oriented towards sustainability, td sustainability sciences appeal to problem-solving as their sole target. At the same time, they are characterised by a critical stance towards modern institutionalised demarcations and understandings of research, such as scientific objectivity and progress.

This contribution highlights epistemologies of the problematic for td sustainability research against the background of the problematic constitution of the hegemonic discourse of SD as a critical, problematising discourse-analytical approach towards problems in td sustainability sciences. The hegemonic discourse of SD is aligned to a neoliberal economic-political interpretation of organising a modern way of life (Castree 2002). Sustainability, thus, under the roof of SD, might serve to strengthen and spread neoliberal hegemony and is the product of a culture, based on a 'Eurocentric Cartesian worldview' (Vanhulst/Beling 2014: 59; Meyer/Vilsmaier 2020), that has a specific relationship, namely a separating, between the individual and the collective, humanity and nature (Moore 2016). Ecological interests, as well as the concern that 'no one will be left behind' (SDGs 2015: Preamble) are then left in marginalised positions. The hegemonic discourse on SD likewise requires the unsustainability of modern ways of life and economy (Dingler 2003) and does not deal with the unsustainable state of the national economy in transnational markets (Parenti 2016).

Thus, the project of td sustainability research offers a problematic opportunity for its own restructuring. A sustainability (research) ethics of the problematic will on the one hand react to (historically conditioned) dependencies and asymmetries (such as hegemony) (Harrasser/Sohldju 2016; Acosta/Brand 2018), thus recognising a true materialistic core of sustainability. But on the other hand be dynamic – as a backwardness to the dynamics of life itself – and open. Then, td sustainability research engages with its problematic of hegemonic structures in science, characterised by a 'double normative framing' (Kläy et al. 2015), founded in liberalism itself. But the problematic is just as well a force to initiate a transdisciplinary and ethical way of relations between entities, which can unfold according to the hegemonic conditions. Reviewing Judith Shklar's 'Liberalism of Fear', Seyla Benhabib and Hannes Bajohr write that we will have to ultimately draw 'a clear line between liberal market capitalism and the political essence of liberalism' (Benhabib 2013: 67, my translation), namely the 'ability to place oneself in the position of the victims' (Bajohr 2013: 145, my translation). In terms of td sustainability research, this means engaging 'not in the back but in the face' (Harrasser/Sohldju 2016: 86, my translation) of social change (Meckesheimer 2013), and 'studying with, and not only about social groups, or at least studying the hegemonic articulations of power' (Mato 2000), namely of ourselves (Freire 2007 [1968]; Fals Borda 1995).

Problems in the context of SD are conceptualised and essentialised differently, as the UN conferences, based on the need to decide between poverty reduction and environmental protection, show. This, in turn, testifies to their relative momentariness. Sustainability thus demands a problem definition of a case-based singularity (Maniglier 2019), in which the internal and external references in the way of individually becoming are recognised, shaped and assessed. Td sustainability research can therefore be understood as complex insofar as we see ourselves as part of the problem (van der Leeuw/Zhang 2014) and do not confront a research topic as a problem. If we reinforce this research paradigm, td sustainability research can process the interweaving of epistemological and normative dimensions. Further work towards epistemologies of the problematic, and a sustainable future, ways of life and cosmologies, beyond the European, should be explored against the background of European perspectives and theories on the concept of the problematic (Vanhulst/Beling 2014; Maniglier 2019).

References

Acosta, Alberto/Brand, Ulrich (2018): Radikale Alternativen: Warum man den Kapitalismus nur mit vereinten Kräften überwinden kann, München: oekom.
Aicher, Otl (1991): Die Welt als Entwurf, Berlin: Ernst & Sohn.

Albán, Adolfo/Rosero, José (2016): 'Coloniality of the natural world: imposition of technology and epistemic usurpation? Interculturality, development and re-existence.' In: Nómada, 45, pp. 27-41.

Bachelard, Gaston (2012): 'Corrationalism and the problematic.' In: Radical Philosophy 173, pp. 27-32.

Bajohr, Hannes (2013): 'Am Leben zu sein heißt Furcht zu haben: Judith Shklar's negative Anthropologie des Liberalismus.' In: Bajohr, Hannes (ed.), Der Liberalismus der Furcht, Berlin: Matthes & Seitz, pp. 131-168.

Benhabib, Seyla (2013): 'Judith Shklar's dystopischer Liberalismus.' In: Bajohr, Hannes (ed.), Der Liberalismus der Furcht, Berlin: Matthes & Seitz, pp. 67-86.

Blanchon, David/Graefe, Olivier (2012): 'Radical political ecology and water in Khartoum: A theoretical approach that goes beyond the case study.' In: L'Espace géographique 1/41, pp. 36-50. https://doi.org/10.3917/eg.411.0035

Bowden, Sean/Kelly, Mark (2018): 'Problematizing problems.' In: Angelaki 23/2, pp. 2-7. https://doi.org/10.1080/0969725X.2018.1451457

Brand, Ulrich/Wissen, Markus (2017): Imperiale Lebensweise: Zur Ausbeutung von Mensch und Natur im Globalen Kapitalismus, München: oekom. https://doi.org/10.3726/JP2017.21

Brandner, Vera (2020): Generative Bildarbeit. Zum transformativen Potential fotografischer Praxis, Bielefeld: transcript. https://doi.org/10.14361/9783839450086

Brown, Trent (2016): 'Sustainability as Empty Signifier: Its Rise, Fall, and Radical Potential.' In: Antipode 48/1, pp. 115-133. https://doi.org/10.1111/anti.12164

Bührmann, Andrea/Schneider, Werner (2008): Vom Diskurs zum Dispositiv: Eine Einführung in die Dispositivanalyse, Bielefeld: transcript. https://doi.org/10.14361/9783839408186

Carnau, Peter (2011): Nachhaltigkeitsethik. Normativer Gestaltungsansatz für eine global zukunftsfähige Entwicklung in Theorie und Praxis, München: Rainer Hampp Verlag.

Castree, Noel (2002): 'False Antitheses? Marxism, Nature and Actor-Networks.' In: Antipode 34/1, pp. 111-146. https://doi.org/10.1111/1467-8330.00228

Clarke, Adele (2012): Situationsanalyse: Grounded Theory nach dem Postmodern Turn, Heidelberg: Springer.

Defert, Daniel/Ewald, Francois (2005): Michel Foucault, Schriften in vier Bänden, Dits et Ecrits, Band IV, 1980-1988, Frankfurt/Main: Suhrkamp.

Dingler, Johannes (2003): Postmoderne und Nachhaltigkeit: Eine diskursttheoretische Analyse der sozialen Konstruktionen von nachhaltiger Entwicklung, München: oekom.

Dow, Sheila (2007): 'Variety of Methodological Approach in Economics.' In: Journal of Economic Surveys 21/3, pp. 447-465. https://doi.org/10.1111/j.1467-6419.2007.00510.x

Eblinghaus, Helga/Stickler, Armin (1996): Nachhaltigkeit und Macht: Zur Kritik von Sustainable Development, Frankfurt: IKO-Verl. für Interkulturelle Kommunikation.

Engbers, Moritz (2020): Kultur und Differenz: Transdisziplinäre Nachhaltigkeitsforschung gestalten (Wahrnehmungsgeographische Studien 29), Oldenburg: BIS-Verlag der Carl von Ossietzky Universität Oldenburg.

Enquete-Kommission des 13. Deutschen Bundestages (1997) Schutz der Menschen und der Umwelt: Konzept Nachhaltigkeit, Fundamente für die Gesellschaft von morgen.

Escobar, Arturo (1995): Encountering Development: The Making and Unmaking of the Third World, Princeton: Princeton University Press.

Fals Borda, Orlando (1995): Research for social justice: Some North-South convergences, Plenary Address at the Southern Sociological Society Meeting, Atlanta, April 8. http://comm-org.wisc.edu/si/falsborda.htm (accessed July 22, 2018).

Foucault, Michel (1978): Dispositive der Macht: Michel Foucault über Sexualität, Wissen und Wahrheit, Berlin: Merve.

Freeth, Rebecca/Drimie, Scott (2016): 'Participatory Scenario Planning: From Scenario "Stakeholders" to Scenario "Owners".' In: Environment: Science and Policy for Sustainable Development 58/4, pp. 32-43. https://doi.org/10.1080/00139157.2016.1186441

Freire, Paolo (2007 [1968]): Pedagogy of the Oppressed, New York: Continuum.

Funtowicz Silvio/Ravetz Jerome (1993): 'Science for the post-normal age.' In: Futures 25/7, pp. 739-755. https://doi.org/10.1016/0016-3287(93)90022-L

Gibbons, Michael et al. (1994): The New Production of Knowledge: The Dynamics of Science and Research in Contemporary Societies. London: SAGE.

Gibbons, Michael (1999): 'Science's new social contract with society.' In: Nature 402, pp. C81-C84. https://doi.org/10.1038/35011576

Grunwald, Armin (2011): Conflict-resolution in the Context of Sustainable Development: Naturalistic versus Culturalistic Approaches, Karlsruhe: KIT Scientific Publishing.

Hajer, Maarten (1995): The Politics of Environmental Discourse: Ecological Modernization and the Policy Process, Oxford: Oxford University Press.

Harrasser, Karin/Solhdju, Katrin (2016): 'Wirksamkeit verpflichtet – Herausforderungen einer Ökologie der Praktiken.' In: ZfM 14/1, pp. 72-86.

Hauff, Volker (1987): Unsere gemeinsame Zukunft: Der Brundtland-Bericht der Weltkommission für Umwelt und Entwicklung, Greven: Eggenkamp.

Haug, Frigga (1999): Vorlesungen zur Einführung in die Erinnerungsarbeit, Hamburg: Argument-Verlag.

Herrera, Amilcar et al. (1976): Catastrophe or New Society?: A Latin American World Model, Ottawa: International Development Research Centre.

Höhler, Sabine/Luks, Fred (2004): Die ökonomische Konstruktion ökologischer Wirklichkeit, NEDS Working Papers Nr. 5.

Hopwood, Bill et al. (2005): 'Sustainable development: mapping different approaches.' Sustainable Development 13, pp. 38-52. https://doi.org/10.1002/sd.244

Jerneck, Anne et al. (2011): 'Structuring sustainability science.' In: Sustainability Science 6/1, pp. 69-82. https://doi.org/10.1007/s11625-010-0117-x

Jonas, Wolfgang (2015): 'Social transformation design as a form of research through design (RTD): Some historical, theoretical, and methodological remarks' In: Jonas et al. (eds.), Transformation Design Perspectives on a New Design Attitude, Berlin: De Gruyter, pp. 114-133. https://doi.org/10.1515/9783035606539-010

Kagan, Sacha (2015): 'Artistic research and climate science: transdisciplinary learning and spaces of possibilities.' In: JCOM: The Journal of Science Communication 14/1, p. C07. https://doi.org/10.22323/2.14010307

Kaika, Maria (2003): 'Constructing scarcity and sensationalising water politics: 170 days that shook Athens.' In: Antipode 3/5, pp. 919-954. https://doi.org/10.1111/j.1467-8330.2003.00365.x

Kambites, Carol J. (2014): '"Sustainable Development": the "Unsustainable" Development of a Concept in Political Discourse.' In: Sustainable Development 22/5, pp. 336-348. https://doi.org/10.1002/sd.1552

Kates, Robert et al. (2001): 'Sustainability science.' In: Science 292, pp. 641-642. https://doi.org/10.2139/ssrn.257359

Kläy, Andreas/Zimmermann, Wolfgang A./Schneider, Flurina (2015): 'Rethinking science for sustainable development: Reflexive interaction for a paradigm transformation.' In: Futures 65, pp. 72-85. https://doi.org/10.1016/j.futures.2014.10.012

Krohn, Wolfgang/Grunwald, Armin/Ukowitz, Martina (2017): 'Transdisziplinäre Forschung revisited: Erkenntnisinteresse, Forschungsgegenstände, Wissensform und Methodologie.' In: GAiA 26/4, pp. 341-347.https://doi.org/10.14512/gaia.26.4.11

Lury, Celia (2020): 'Compositional methodology: on the individuation of a problematic of the contemporary.' In: Leistert, Oliver/Schrickel, Isabell (eds.), Thinking the Problematic. Genealogies and Explorations between Philosophy and the Sciences, Bielefeld: transcript.

Maniglier, Patrice (2019): 'Problem and Structure: Bachelard, Deleuze and Transdisciplinarity.' In: Theory, Culture & Society, online July 15th. https://doi.org/10.1177/0263276419878245

Maniglier, Patrice (2012): 'What is a problematic?' In: Radical Philosophy 173, pp. 21-23.

Mato, Daniel (2000): 'Studying with the Subaltern, not only about the Subaltern Social Groups, or, at Least, Studying the Hegemonic Articulations of Power.' In: Nepantla (ed.), Views from South 1.3, Durham, NC: Duke University Press, pp. 479-503.

Meadows, Donella et al. (1972): The Limits to Growth: A Report for the Club of Rome's Project on the Predicament of Mankind, New York: Universe Books. https://doi.org/10.1349/ddlp.1

Meckesheimer, Anika (2013): 'Decolonization of social research practice in Latin America: What can we learn for German social sciences?' In: Transcience 4/2, pp. 79-98.

Meyer, Esther (2020): 'Solvable problems or problematic solvability? Problem conceptualization in transdisciplinary sustainability research and a possible epistemological contribution.' In: GAiA 29/1, pp. 34-39. https://doi.org/10.14512/gaia.29.1.8

Meyer, Esther/Peukert, Daniela (2020): 'Designing a Transformative Epistemology of the Problematic: A Perspective for Transdisciplinary Sustainability Research.' Social Epistemology, online. https://doi.org/10.1080/02691728.2019.1706119

Meyer, Esther/Vilsmaier, Ulli (2020): 'Economistic discourses of sustainability: determining moments and the question of alternatives.' In: Sustainability in Debate 11/1, pp. 98-110. https://doi.org/10.18472/SustDeb.v11n1.2020.26663

Michelsen, Gerd/Adomßent, Michael (2014): 'Nachhaltige Entwicklung: Hintergründe und Zusammenhänge.' In: Heinrichs, Harald/Michelsen, Gerd (eds.), Nachhaltigkeitswissenschaften, Berlin/Heidelberg: Springer Spektrum, pp. 3-60. https://doi.org/10.1007/978-3-662-44643-0_1

Moncayo, Víctor (2009): Fals Borda, Orlando, 1925-2008: Una sociología sentipensante para América Latina, Bogotá: Siglo del Hombre Editores/CLACSO.

Moore, Jason (ed.) (2016): Anthropocene or Capitalocene? Nature, History, and the Crisis of Capitalism, Oakland: PM Press.

Nowotny, Helga et al. (2001): Re-thinking Science: Knowledge and the Public in an Age of Uncertainty, Cambridge, UK: Polity Press.

Olssen, Mark (2014): 'Discourse, Complexity, Normativity: Tracing the elaboration of Foucault's materialist concept of discourse.' In: Open Review of Educational Research 1/1, pp. 28-55. https://doi.org/10.1080/23265507.2014.964296

Osborne, Peter (2015): 'Problematizing Disciplinarity, Transdisciplinary Problematics.' In: Theory, Culture & Society 32/5-6, pp. 1-33. https://doi.org/10.1177/0263276415592245

Ott, Konrad (2014): 'Institutionalizing Strong Sustainability: A Rawlsian Perspective.' In: Sustainability 6, pp. 894-912. https://doi.org/10.3390/su6020894

Ott, Konrad/Döring, Ralf (2004): Theorie und Praxis starker Nachhaltigkeit, Marburg: Metropolis.

Parenti, Christian (2016): 'Environment-Making in the Capitalocene: Political Ecology of the State.' In: Jason Moore (ed.), Anthropocene or Capitalocene? Nature, History, and the Crisis of Capitalism, Oakland: PM press, pp. 166-184.

Peukert, Daniela/Vilsmaier, Ulli (2019): 'Entwurfsbasierte Interventionen in der transdisziplinären Forschung.' In: Ukowitz, Martina/Hübner, Renate (eds.), Wege der Vermittlung – Intervention – Partizipation, Wiesbaden: Springer-Verlag. https://doi.org/10.1007/978-3-658-22048-8_10

Richter, Ralph/Rohnstock, Nepomuk (2016): 'Der Erzählsalon als Verfahren strategischen Erzählens: Konzeptionelle Voraussetzungen und empirische Gestalt am Beispiel des Projektes Lausitz an einen Tisch.' In: DIEGESIS: Interdisciplinary E-Journal for Narrative Research 5/2, pp. 84-100.

Rockström, Johan et al. (2009): 'Planetary boundaries: exploring the safe operating space for humanity.' In: Ecology and Society 14/2, p. 32. https://doi.org/10.5751/ES-03180-140232

Sachs, Wolfgang (2002): Nach uns die Zukunft: Der globale Konflikt um Gerechtigkeit und Ökologie, Frankfurt/Main: Brandes & Apsel.

Sachs, Wolfgang (1993): Wie im Westen so auf Erden: Ein polemisches Handbuch zur Entwicklungspolitik, Reinbek: Rowohlt.

Satanarachchi, Niranji/Mino, Takashi (2014): 'A framework to observe and evaluate the sustainability of human-natural systems in a complex dynamic context.' In: SpringerPlus 3/618. https://doi.org/10.1186/2193-1801-3-618

Savransky, Martin (2018): 'The Humor of The Problematic: Thinking with Stengers.' In: SubStance 57/1, pp. 29-46.

Schmidt, Jan C. (2011): 'What is a problem? On problem-oriented interdisciplinarity.' In: Poiesis Prax 7, pp. 249-274. https://doi.org/10.1007/s10202-011-0091-0

Schmieg, Gregor et al. (2018): 'Modeling normativity in sustainability: A comparison of the Sustainable Development Goals, the Paris Agreement, and the Papal Encyclical.' In: Sustainability Science 13/3, pp. 785-796. https://doi.org/10.1007/s11625-017-0504-7

Scott, David (2014): Gilbert Simondon's Psychic and Collective Individuation: A Critical Introduction and Guide, Edinburgh: Edinburgh University Press.

Simondon, Gilbert (2007): 'Das Individuum und seine Genese.' In: Blümle, Claudia/Schäfer, Armin (eds.), Struktur, Figur, Kontur: Abstraktion in Kunst und Lebenswissenschaften, Zürich: diaphanes, pp. 29-46.

Sneddon, Chris et al. (2006): 'Sustainable development in a post-Brundtland world.' In: Ecological Economics 57, pp. 253-268. https://doi.org/10.1016/j.ecolecon.2005.04.013

Spangenberg, Hans Joachim (2011): 'Sustainability science: a review, an analysis and some empirical lessons.' In: Environmental Conservation 38/3, pp. 275-287. https://doi.org/10.1017/S0376892911000270

Stengers, Isabelle (2005): 'Introductory Notes on an Ecology of Practices.' In: Cultural Studies Review 11/1, pp. 183-196. https://doi.org/10.5130/csr.v11i1.3459

Stengers, Isabelle (2010): Cosmopolitics I, Minneapolis: University of Minnesota Press.

Strong, LaToya et al. (2016): 'Against Neoliberal Enclosure: Using a Critical Transdisciplinary Approach in Science Teaching and Learning.' In: Mind, Culture, and Activity 23/3, pp. 225-236. https://doi.org/10.1080/10749039.2016.1202982

Thompson Klein, Julie (2014): 'Discourses of transdisciplinarity: Looking back to the future.' In: Futures 63, pp. 68-74. https://doi.org/10.1016/j.futures.2014.08.008

Thompson Klein, Julie et al. (2001): Transdisciplinarity: Joint Problem Solving among Science, Technology, and Society: An Effective Way of Managing Complexity. Basel: Birkhäuser.

Timpf, Siegfried (2000): Das Dispositiv der zukunftsfähigen Entwicklung, Hamburg: Doctoral thesis HWP.

Transforming our world: the 2030 Agenda for Sustainable Development (SDGs), Resolution, A/RES/70/1.

Van der Leeuw, Sander/Zhang, Yongsheng (2014): 'Are We Part of the Solution or Part of the Problem?' SantaFe Institute, Working Paper.

Vanhulst, Julien/Beling, Adrián. (2014): 'Buen vivir: Emergent discourse within or beyond sustainable development?' In: Ecological Economics 101, pp. 54-63. https://doi.org/10.1016/j.ecolecon.2014.02.017

Vanhulst, Julien/Hevia, Antonio (2016): 'Los Senderos Bifurcados del Desarrollo Sostenible: Un Analísis del Discurso Académico en América Latina.' In: Floriani, Dimas/Helvia, Antonio (eds.), América Latina: Sociedade e Meio Ambiente. Teorias, Retóricas e Conflitos em Desemvolvimento, Curitiba: UFPR.

Vanhulst, Julien/Zaccai, Edwin (2016): 'Sustainability in Latin America: An analysis of the academic discursive field.' In: Environmental Development 20, pp. 68-82. https://doi.org/10.1016/j.envdev.2016.10.005

Vilsmaier, Ulli et al. (2015): 'Case-based Mutual Learning Sessions: knowledge integration and transfer in transdisciplinary processes.' In: Sustainability Science 10/4, pp. 563-580. https://doi.org/10.1007/s11625-015-0335-3

Vilsmaier, Ulli et al. (2017): 'Research in-between: The constitutive role of cultural differences in transdisciplinarity.' In: Transdisciplinary Journal of Engineering & Science 8, 169-179. https://doi.org/10.22545/2017/00093

Vilsmaier, Ulli (2018): 'Grenzarbeit in integrativer und grenzüberschreitender Forschung.' In: Heintel, Martin et al. (eds.), Grenzen: Theoretische, konzeptionelle und praxisbezogene Fragestellungen zu Grenzen und deren Überschreitungen, Wiesbaden: Springer VS, pp. 113-134. https://doi.org/10.1007/978-3-658-18433-9_6

Voss, Daniela (2018): 'Simondon on the notion of the problem.' In: Angelaki 23/2, pp. 94-112. https://doi.org/10.1080/0969725X.2018.1451471

Williams, Colin/Millington, Andrew (2004): 'The diverse and contested meanings of sustainable development.' In: The Geographical Journal 170, pp. 99-104. https://doi.org/10.1111/j.0016-7398.2004.00111.x

World Commission on Environment and Development (1987): Our Common Future, Oxford: Oxford University Press.

Ziegler, Rafael/Ott, Konrad (2015): 'The quality of sustainability science: A philosophical perspective.' In: Enders, Judith/Remig, Moritz (eds.), Theories of Sustainable Development, London/New York: Routledge, pp. 43-64. https://doi.org/10.5771/9783845258430-15

A Genealogical Perspective on the Problematic: From Jacques Martin to Louis Althusser

Jean-Baptiste Vuillerod

The current importance of the notion of the problematic invites us to think about its relevance and its conceptual content, but also to explore its genealogy in the works that explicitly refer to the problematic as a philosophical concept. Thus, it is often considered that the word 'problématique' appears for the first time in 1949 in Gaston Bachelard's *Le rationalisme appliqué* (Bachelard 2004: 51). Even the statistical studies cannot find an occurrence of this word before 1949 and its apparition in Bachelard's and Paul Ricoeur's works (Benoit 2005). Its conceptual signification goes back further in France, however. Indeed, the introduction of a manuscript dating from 1947 and entitled 'Some remarks on the notion of the individual in Hegel's philosophy' was entirely devoted to the thought of the problematic. Nowadays located in Louis Althusser's archives, this manuscript is nothing but the master thesis that Jacques Martin, a nearly unknown student, wrote about Hegel under the direction – that is to be noticed – of Bachelard.

We know very little about Martin, except that Althusser owes him the notion of the problematic. Thanks to Yann Moulier Boutang's work, we know that he was born May 18, 1922 in Paris and joined the École Normale Supérieure in 1941 (Moulier Boutang 2002: 376-393). There he became one of Althusser and Michel Foucault's closest friends. Martin was a very brilliant student, passionate about German philosophy and notably Hegel and Marx. But he also suffered from depression and mental illness, which led him to inactivity and finally suicide. This is the reason why, as Nikki Moore insisted (Moore 2005), he is the 'man without work' that Foucault references in *History of Madness* and in *Madness, the Absence of Work* (Foucault 1995; 2006). His master's thesis was never published during his lifetime and Martin just translated some of Hegel's, Wiechert's and Hesse's work (Hegel 1948; Wiechert 1953; Hesse 1955).

The name of Jacques Martin was therefore apparently destined to fall into oblivion. Fortunately, Althusser preserved the master's thesis and recognized his debt towards him in *For Marx*, mentioning Martin as the real inventor of the concept of the problematic: 'I thought it possible to borrow for this purpose the concept of a *"problematic"* from Jacques Martin to designate the particular unity of a theoretical formation and hence the location to be assigned to this specific difference [...]' (Althusser 1969: 32). After Martin committed suicide in 1963, Althusser was profoundly shocked and this is the reason why he dedicated *For Marx* to him, the person that led him to the reading of Marx: 'These pages are dedicated to the memory of Jacques Martin, the friend who, in the most terrible ordeal, alone discovered the road to Marx's philosophy – and guided me onto it.'

Nowadays Jacques Martin's text is published at last (Martin 2020) and we can evaluate the real significance of his reflexion. The early development of the notion of a problematic in Martin's work and its importance for Althusser, one of the most famous and strongest proponents of the concept, calls into question the traditional genealogy of the notion and means that a new genealogical perspective on the problematic has to be pursued. Our objective here is to contribute to this debate by analyzing the intellectual context in which Martin used the word as a philosophical concept, then in presenting the signification of the problematic in Martin's view, and finally in confronting Althusser's and Martin's comprehensions of this notion to set out the philosophical issues of this genealogical perspective.

From Germany to France

When Martin wrote his master's thesis in 1947, the problematic was not yet designated as a philosophical concept in France and was not considered as a powerful and relevant tool for analysis. But the word existed in the French intellectual area, especially in the philosophy of science. It is likely that the word had been imported from Germany, where Heidegger made a specific use of it and tried to provide a rigorous concept of the problematic in discussing Hartmann's and Windelband's works. Thus Martin picked up the problematic at the crossroads of those different influences and made it his own to turn it into a very specific concept destined to have a great posterity in French philosophy. Let us sum up briefly here those steps that preceded Martin's appropriation of the notion.

In the 1923-1934 lessons that he gives in Marburg – now published and entitled *Einführung in die Phänomenologische Forschung* (Heidegger 1994) – Heidegger proposes an intense reflection about the notion of problem and its implications for the history of philosophy. In section 10, he notably deals with the 'clarification of problems (*Klärung der Probleme*)' and distinguishes between problems and questions. A question refers to an implicit care of the *Dasein*: '*Suchen als eine bestimmte Sorge des Daseins*' (Heidegger 1994: 73). In Husserl's philosophy, discussed by Heidegger, it is the question of knowledge that is crucial, conceived as a care of an absolute clarity, a sake of clearness (Heidegger 1994: 79). But a question is not exactly a problem, in the sense that a problem is the question that is explicitly stated and raised in an explicit way (Heidegger 1994: 73). The question appears as the opening of the *Dasein* into the beings and the care that founds such an opening. But the problem is the explicitation of this ontological state of the *Dasein* as a necessary attitude towards the worlds (Heidegger 1994: 77). In 1923-24, the word 'problematic' does not appear yet, but Heidegger insists on the importance of the *Fragestellung* – the question stating – as part of the process of making the question explicit.

In *Einführung in die Phänomenologische Forschung*, Heidegger refers to the School of Marburg, to Wilhelm Windelband and to Nicolaï Hartmann's philosophies as important reflections about the notion of the problem and its application to the history of philosophy. In *Zur Methode der Philosophiegeschichte*, written in 1909, Hartmann tried to understand the history of philosophy through the notion of problem (Hartmann 1958). According to him, problems would be the only way to overtake the idiosyncrasy of thinkers and to restore continuity in the history of thought. The problems are transmitted through the ages and thinkers progress in their resolutions. Thus Hartmann criticized Windelband, who had already conceived the history of philosophy as a history of problems, but who considered that those problems were not independent from the living and cultural conditions of the authors (Windelband 1912). On the contrary, in Hartmann's view, there is no link between the history of problems and the history of thinkers. It does not matter if sometimes the philosophers do not resolve the problems or change their preoccupations or ignore them, because the next ones will take care of them and will pursue the task of resolving them.

Heidegger indicates that those comprehensions of the history of philosophy through the notion of a problem were a 'starting point (*Standpunkt*)' for his own research (Heidegger 1994: 78), but that they had to be overcome by an ontological perspective, stepping forward to the source of the question (*Quellen und Motive des Fragens*): the *Dasein* itself. Indeed, for him, the problem has to reveal the question that is at the root of its existence. In 1927, with the publication of *Sein und Zeit*, Heidegger gives a conceptual name to the *Fragestellung*: calling it the *Problematik*. The book opens on the oblivion of the question about being as 'a thematised question of a real research (*als thematische Frage wirklicher Untersuchung*)' (Heidegger 1977: 3). In this context, Heidegger uses the word problematic to point out the renewal of the question of being and the possibility of an explicit reflection about it. He speaks about 'the possibility of reaching an ontological founded problematic (*die Möglichkeit der Inangriffnahme einer zureichend fundierten ontologischen Problematik*)' (Heidegger 1977: 18). Therefore the problematic appears as the new philosophical term for the stating of the question.

Although the importance of the notion of the problematic in 'Sein und Zeit' is obvious, it seems that it was not this book that introduced the word in France, but another text that was translated long before and popularized Heidegger's thought on a large scale (Janicaud 2001: 40): *Vom Wesen des Grundes*, written in 1929 (Heidegger 1976) and published in France as early as 1938 thanks to Henry Corbin's translation (Heidegger 1968). In this text, Heidegger wants to bring to light what he calls 'the ontological problematic' (Heidegger 1968: 100), the 'problem of Being', which was 'repressed' by the tradition, but nevertheless was always present implicitly (Heidegger 1968: 156). The task Heidegger assigns to his own philosophy is to put forward a 'problematic explicitly worded of the concept of Being' (Heidegger 1968: 98). As in Hartmann and Windelband, the problematic is here connected to the history of philosophy, but conversely, Heidegger conceives the problematic as the unique question that underlies all the history of thought. Such a question is thought of as the ontological difference and, according to Heidegger, it is the task of philosophy to make the ontological difference explicit in a clear problematic.

This formulation of a precise concept of the problematic in Germany, particularly in Heidegger's work, is important to understanding the discussion around the concept in France during the 1930s and 1940s. Indeed, we have reason to believe that the text *Vom Wesen des Grundes* was central in the discussion between Jean Cavaillès and Albert Lautman in the *Société française de philosophie* in February 1939.[1] In his conference presentation, Lautman refers twice to Heidegger (Cavaillès 1994: 608, 630) and the notions of problem and problematic are at the center of the discussion. It is in light of the dialectical concept that the problematic is understood by Lautman and Cavaillès. Indeed, Cavaillès calls the 'fundamental dialectic of mathematics' the dynamic process of problem solving: 'It could be called the fundamental dialectic of mathematics: if the new notions appear as required by the given problems' (Cavaillès 1994: 601). In a Hegelian perspective, he understands the problematical dialectic as a historical process that goes forwards from problems to solutions. For his part, Lautman uses the term 'problematic' and claims that 'dialectics in itself is pure problematic' (Cavaillès 1994: 607). According to him, dialectics is the science of philosophical and abstract problems that are not mathematical (for example, the problem of essence and existence, of matter and form, of finite and infinite, and so on), and mathematical notions are answers to those metaphysical problems. Like Cavaillès, Lautman identifies the problematic with dialectics. But like Heidegger, in a Platonic tradition, he removes the dialectical problematic from a concrete history and considers that they cross through all the history of philosophy as transversal transcendent ideas. In the discussion, Jean Hyppolite stresses the difference between Cavaillès and Lautman in their discussions of dialectics and takes the side of a Hegelian concept of dialectic in which problems change at the same time as history moves forward (Cavaillès 1994: 619-620). But both make extensive use of the notions of problem and introduce the term of the problematic taken from Heidegger in France.

When Jacques Martin developed his own conception of the problematic in a historical and Hegelian way, he was probably aware of this discussion and he picked up the term from this epistemological appropriation coming from Heidegger's notion. We can now move on to the very specific meaning that Martin gives to the problematic.

---

1 Concerning Lautman, Emmanuel Barot explains that he had read *Vom Wesen des Grundes* and that he had appropriated the notions of this essay (Barot 2009: 138-144).

Martin's problematic

The meaning of the problematic in Martin's text and the importance of this notion for him can only be understood by analyzing the particular goal that he sets for himself in his master's thesis: reading Hegel through the lens of Marx and, more precisely, finding in Hegel a precursor of the Marxist criticism of the bourgeois individual. Martin explicitly considers that the two authors enlighten each other in the way of thinking about the relationship between the individual and her social and historical determination: 'Thinking history as something effective (*dire que l'histoire est effective*) means that, on the philosophical level, Hegel's philosophy was an object of critical reflection for Marx; it is only in reference to this one that the indications of Hegel about the individual can be appreciated [...].' (Martin 2020: 41)

It is in this context that Martin uses the notion of the problematic. If the problematic is required by the Marxist reading of Hegel that Martin proposes in 1947, it is because he has to justify why he may raise the problem of the individual in Hegel's philosophy even though Hegel did not thematize it explicitly and, consequently, did not address it in a direct way. Indeed Martin needs a notion that could indicate the possibility of reading the history of philosophy in revealing some implicit questions in Hegel's thought. As Martin recognizes, 'the problem of the individual was not addressed in Hegel's philosophy (Martin 2020: 39). But this is precisely the reason why he has to think about a new way of reading Hegel and, therefore, why he has 'to institute a problematic to contribute to locate the importance of those themes'. The institution of the problematic means the elaboration of a particular perspective of reading, in light of a problem raised by the history of philosophy, but that remains implicit in a text. The notion of the problematic can thus resolve the issue of finding a Marxist question in a theory that came before Marx in the history of thought.

Through his original and very specific approach, Martin shifts the notion of the problematic from Heidegger's ontological perspective and from the epistemological debate between Lautman and Cavaillès to the political Marxist field of thought. In doing so, he transforms profoundly the meaning of the concept. The problematic, as developed by Heidegger, but also by Lautman, entailed exactly the means of asking a question to thinkers, even though this question was not explicitly raised in their philosophy. But Martin does not consider that an ontological question might be the unique cross-cutting of all the history of philosophy, and he does not think that some eternal and ideal problems are present in particular mathematical problems. On the contrary, like Cavaillès, he thinks that problems are totally historical and that only history can explain the implicit problematic of a thinker. According to him, a problematic does not transcend history and it is history itself that creates new problematics when it allows us to read the authors of the past in the light of some more recent authors – in this case Hegel in the light of Marx.

Nevertheless, Martin's objective should not be understood as the will of an ideologue plotting to incorporate Hegel in his political and strategic enterprises. Martin does not even join the French Communist Party, even though he shares a lot of their views (Moulier Boutang 2002). If he returns to Hegel from Marx, it is because he thinks that Hegelian concepts can help us to better understand Marx's philosophy itself. Like a lot of Marxists at that time, Martin thinks that Marx had developed a very precious science of history and of the economic conditions of the capitalist world, but had not explicitly exposed the philosophy that supported its explanations, making it difficult to actualize Marxist thought in the new capitalist context of the aftermath of World War II. It is therefore a very pressing task to explain the philosophy of Marx in the light of Hegel. This idea is notably claimed by Althusser in the master's thesis he writes in the same year, in 1947: 'Hegel is Marx's silent rigor, the living truth of a body of thought which is too pressed by circumstances to apprehend itself in self-consciousness, but which betrays itself in the least of its movements' (Althusser 2014: 142). Martin seems to share the same opinion as his friend. Both think that Hegel can provide the philosophy required by Marxism. We can notice that such a project, even after Martin's death, will be the aim of Althusser's life until his last reflections (Althusser 1994).

In terms of content, the problematic enables Martin to inscribe the individual in its social and historical conditions and in this way to criticize the solipsism of the bourgeois conception of the individual. It connects Hegel to Marx's criticism of the bourgeois individual and thus makes him appear as a critical transition between the individualistic thought of the 18th century and its criticism by Marxism: 'Hegel's propositions are nothing if separated from the individualistic conception of the person in Rousseau or Kant – and they are not determined for those who do not read them through Marx's claims, which make possible the meaning of Hegelianism that simultaneously makes Marx possible' (Martin 2020: 44). Thus, for Martin, the problematic is a way to plunge Hegel into the history of thought and to read him as the first critical philosopher of the bourgeois individual.

As Marx, and Hegel before him, would object to the idea of the subject's self-sufficiency, he integrated the individual in society and history and set out all the mediations that contribute to create personality and subjectivity. According to Martin, Hegel had 'an intuition of the individual as integrated in a reality from which he cannot be separated (*une intuition de l'individu comme intégré dans une réalité dont il n'est pas separable*)' (Martin 2020: 70). Hegel wrote about all the historical and social mediations that determine the individual at one particular time: 'Hegel always conceived the concrete in the form of totality: not a totality of qualities or gifts, whose realization could be enough to define individuality, but the totality of the relationships between the individual and the world that defines her and constitutes her, and conversely those that the individual contributes to constitute and to define.' (Martin 2020: 87) Hegel was the first to propose such a conception of a mediated subjectivity opened to the world and defined essentially by its relations and not by itself. This is exactly why his philosophy is important for Marxism. Thinking the mediations as constitutive of subjectivity, it offers a clear articulation between individuals and collectivity that wipes out the solipsistic and individualist approach of man that characterizes the bourgeois point of view.

In such a view, and in the horizon of a comparison between Martin's and Althusser's problematics, we could sum up the comprehension that Martin had of the problematic in three points: the problematic is a question, it is essentially diachronic and it is a way to read together Hegel and Marx.

1. In Martin's view, a problematic is a specific problem, a particular question, a thematic. Martin speaks of 'the theme of the individual (*le thème de l'individu*)' (Martin 2020: 43) and of 'a theme that offers itself as a problem (*un thème qui se propose comme un problème*)' (Martin 2020: 44). It is not a global structure of thought or a way to raise particular problems, but a single particular problem itself. Martin looks forward to Hegel's criticism of individualism in the Age of Enlightenment and considers the problematic as the designation of such a singular question.

2. Martin's problematic is essentially connected to history and is consequently understood from a diachronic point of view. It is a means to escape from the subjectivity of a thinker and from the explicit questions

that are raised by a philosopher in order to reintegrate a philosophical system in the whole process of the history of thought. This is the reason why, according to Martin, the problematic carries out a 'dispossession by history (*dépossession par l'histoire*)' (Martin 2020: 45). By that means, Hegel is deprived of his own intentional work and is questioned with a problematic that belongs to the later history of Marxism.

3. Martin's objective is to promote a reading of Hegel that would be compatible with Marx. His goal is to read Hegel and Marx together thanks to the common problematic of the criticism of the bourgeois individual. In some ways, it is the idea that Marx had not completely developed his philosophy and that Marxism needs a philosophical theory that meets its practical aspirations.

It is only by keeping in mind these elements that we will understand the differences that Althusser introduces to the notion of the problematic in the 1960s. In spite of those differences, Althusser recognizes his debt towards his friend Jacques Martin, who had accomplished the decisive action of shifting the problematic from Heidegger's philosophy and from the French epistemological debates between Lautman and Cavaillès to the Marxist space of thought, and thus had given the impulsion of a new Marxist theory that could be improved thanks to the use of this notion; henceforth it was truly reflected and worked as a legitimate concept.

From Martin to Althusser

My objective here is to analyze the way Althusser inherits the notion of the problematic from his friend Jacques Martin and how by doing this he transforms the notion at the same time in a decisive way. It is only this double movement of inheritance and transformation that can explain how Althusser is able to recognize his debt towards his friend while creating one of the most representative and powerful concepts of French philosophy during the 1960s.

In *For Marx*, Althusser uses the notion of the problematic to reflect on the 'epistemological break (*coupure épistémologique*)' (Althusser 1969: 32) that occurred between Marx and the philosophers that came before him, especially Hegel and Feuerbach (Gillot 2009: 31). Against Hegel's teleological dialectic, focused on the unity of spirit and on the end of history, Marx opposed a complex dialectic that could contain the 'overdetermination' (Althusser 1969: 87) of a singular event by the diversity of all the elements of the society, understood as a 'complex structured whole' (Althusser 1969: 193). And against Feuerbach's humanism, which had referred to an ahistorical human nature, Marx dispensed with the unscientific and ideological concept of Man and replaced it with a scientific view of society and its history, based on social structures in which men were limited to occupying functions (Althusser 1969: 219-241). The date of the break would have been 1845, when Marx wrote with Engels 'The German Ideology' and thus reached a real science of history. Althusser's intentions were perfectly clear: through Hegel, he targeted the simplistic and rigid Stalinist dialectic, and through Feuerbach, he wanted to criticize the humanist Marxism in France and the Soviet Union that followed Stalin's death. Marx's epistemological break was also Althusser's break with the ideological Marxism of his time.[2]

Althusser's objective is to provide an adequate explanation of the social organization and of the revolutionary process. To this end, he has to take into account the extra-economic causalities that traditional Marxism did not consider since it limited the social contradictions to the conflicts between the productive forces and the relations of production, and thereby restricted the revolution to a transformation of the economic basis. Althusser estimates that such a program is clearly unsatisfactory and needs to be completed by the importance of the political, juridical and ideological factors of the revolution. In particular, the Chinese Cultural Revolution and Mao's criticism of Stalinism proved that a society could change in its economic basis and, despite of this transformation, could remain the same from the point of view of its political and ideological domination.

The complex causality that Althusser proposed by reading Marx in a new perspective is precisely dedicated to thinking these pluralistic phenomenon. But he was convinced that this purpose cannot be achieved if we continue to read Marx in the light of the Hegelian legacy. Hegel's philosophy developed a simple, or even a simplistic concept of causality, where each society is structured by a fundamental contradiction and by a unique principle – for example the juridical principle in the Ancient Rome. Regarding this point, Feuerbach did not differ from Hegel, when he considered each social state

---

2 On the context of Althusser's thought, see Elliott 1987 and Lewis 2005.

and the whole history of humanity as constituted by the self-alienation of humankind. In each case, it is the philosophical desire for a first and unique principle that is at stake. It is exactly this Hegelian simplicity, extended by Feuerbach, that can be found in the traditional Marxism, that focuses only on the economic contradiction and ignores other social contradictions. This is the reason why Althusser decides to insist on the epistemological break between Marx and his predecessors and when he reads Martin's text to find a concept that could help him to express this theoretical and historical shift.

The problematic is therefore required to think this epistemological break of Marxism. Althusser gives some dispersed definitional elements that characterize it. According to him, the problematic is 'the constitutive unity of the effective thoughts' (Althusser 1969: 66) of an author; 'the *typical systematic structure* unifying all the elements of the thought' (Althusser 1969: 67); a way by which a philosophy or an ideology reflects its objects, '*the way it reflects that object* (and not in the object itself)' (Althusser 1969: 66); 'the system of *questions* commanding the answers given by the ideology' (Althusser 1969: 67); more generally the 'theoretical presuppositions' (Althusser 1969: 68) of thought; and an unconscious element of thought supposing that 'a philosopher *thinks in it rather than thinking of it*' (Althusser 1969: 69). We can thus say that Althusser considers the problematic as a way of questioning and reflecting objects that provide unity to thought and of which the philosopher is never absolutely conscious.

Hence the problematic describes the organisational mode of a system of thought, the way of thinking it entails, and the particular ways of raising and solving particular questions. In Marx' theory, according to Althusser, it means that the questions are never raised by presupposing a simplistic causality, even if this causality would be the economic contradictions and not the spiritual principle of a society (Hegel) or the alienation of humankind (Feuerbach). The resolutions that he proposes are also not instructed by a single phenomenon. Marx always takes into account the diverse factors that constitute each society and he underlines the multiplicity of causes – the overdetermination – that are at stake in the revolutionary movement. His manner of formulating problems is not the Hegelian way of thinking, and in this sense we can say that he thinks in a different problematic.

On this basis we can understand the difference that has arisen in the concept of problematic between Jacques Martin and Louis Althusser. Althusser's polemical perspective against Hegel's and Feuerbach's philosophies brings him to transform the notion. In his view, it is necessary for the problematic to be understood as a means to separate all the mature thought of Marx from other philosophies. From there stems the differences with Martin and the fact that Althusser considers the problematic not as a single question raised from a diachronic point of view in order to reconcile Hegel and Marx, but as a systematic and synchronic structure of thought absolutely original and revolutionary, without any link to any prior philosophy.

1. In Althusser's discourse, the problematic does not concern a thematic or a unique question, but a whole organization of thought from which the particular questions can be raised. Althusser deals with 'the active but unavowed problematic which fixes for it the meaning and movement of its problems and thereby of their solutions' (Althusser 1969: 69). This means that the singular elements of thought should be considered from the problematic and not the opposite: 'So anyone who still wants to pose the problem of elements in this perspective must recognize that everything depends on a question which must have priority over them: the question of the nature of the problematic which is the starting-point for actually thinking them, in a given text.' (Althusser 1969: 68) Or, as he also writes: 'Every ideology must be regarded as a real whole, internally unified by its own problematic, so that it is impossible to extract one element without altering its meaning.' (Althusser 1969: 62) Thus the problematic is not a problem, but a way or a perspective to raise problems. It is not a particular question – Martin's question of the individual – but a principle of coherence between all the questions that a philosophy can ask. Feuerbach's problematic, for example, was anthropology, a way of questioning from the presupposition of human nature and from the point of view of human relationships. On the contrary, Marx discovered a problematic where social structures and structural relations, not men, were at the center.

2. Being a structure of thought and not a particular question, the problematic is set out by Althusser from a spatial figure and not from a temporal perspective. It is therefore not understood as diachronic, as it was in Martin, but as synchronic. According to Althusser, the problematic is a 'field' (Althusser 1969: 66) and it is not constituted by the succession of thoughts in history, but by the combination of different elements inherent to a philosophy. From this principle, reading Hegel in light of the individualistic thought of the 18th

century and of Marx is not relevant; much more so is searching in Marx's work for the moment – after 1845 – when his thought found a new systematic way of raising problems and hence became absolutely original.

3. The consequence is that Althusser does not want to read Hegel and Marx together but, on the contrary, strives to distinguish Marx from all the philosophies that preceded his, especially from Hegel's dialectic and Feuerbach's humanism. He searches for what is called, in an Aristotelian language, the 'specific difference' that separates Marx from others, and then defines the problematic as 'the particular unity of a theoretical formation and hence the location to be assigned to this specific difference' (Althusser 1969: 32). The objective cannot be, as it was in Martin, to reconcile Marx and his predecessors, but to inscribe the very originality of Marx in the history of philosophy: 'The truly Marxist critique of Hegel depends precisely on this change of elements, that is, on the abandonment of the philosophical problematic whose recalcitrant prisoner Feuerbach remained.' (Althusser 1969: 48)

Conclusion

When Bachelard used the word 'problematic' in *Le rationalisme appliqué*, he may have borrowed it from the epistemological debate between Cavaillès and Lautman, but he also could have found it in the work of his young student Jacques Martin, whose master thesis he supervised. In any event, the role of Martin in the development of French philosophy after World War II should surely be revalorized. His importance in the thought of the problematic is enough to reconsider his name in the great tradition of Gilles Deleuze, Michel Foucault, Jacques Derrida, Jean-François Lyotard and others. But his thought of social mediations is not unconnected with the idea of a historical transcendental that is thematized at the end of Althusser's work in 1947 – 'Marx understood that transcendental was history' (Althusser 2014: 170) – and that is exposed for itself in Michel Foucault's 'La constitution d'un transcendental dans la *Phénoménologie de l'esprit* de Hegel'[3], his master's thesis written in 1949. One can also notice that Martin mentions already the psychoanalytic concept of 'overdetermination (*surdétermination*)' (Martin

---

3 Manuscript not yet published.

1947: 31)[4] to explain the complex causality that results from interfering social mediations. And as we said earlier, Foucault's 'absence of work' is a reference to Martin's madness.

My objective here is not to claim that all the French philosophy in the 1960s was contained in Martin's first work. I just would like to sketch the possibility of considering him as an essential moment of its development from the 1940s to the 1960s and even afterwards. The transformation of the notion of the problematic by Althusser is remarkable on this point, because he uses Martin's work but he turns it into something else and in a different context of thought. Martin developed some decisive intuitions and some important concepts that have not been developed further in their original form, but that have been changed to serve a different goal and to signify different meanings. He is one of the links in this philosophical chain that runs to us and we probably would not reflect today on the problematic without his contribution.

References

Althusser, Louis (1965): For Marx, London: The Penguin Press.
Althusser, Louis (1994): Sur la philosophie, Paris: Gallimard.
Althusser, Louis (2014): The Spectre of Hegel: Early Writings, London/New York: Verso.
Bachelard, Gaston (1949): Le rationalisme appliqué, Paris: PUF.
Barot, Emmanuel (2009): Lautman, Paris: Les Belles Lettres. https://doi.org/10.14375/NP.9782251760636
Benoit, Jean-Pierre (2005): 'L'émergence des "mots de la problématisation" dans les sources universitaires et les dictionnaires spécialisés et généraux: constats et analyses.' In: Les Sciences de l'éducation – Pour l'Ère nouvelle 38/3.
Cavaillès, Jean (1994): Œuvres complètes de philosophie des sciences, Paris: Hermann.

---

4 'But this reciprocity implies a dialectic – it could be said provisionally an *overdetermination* of significations – that Hegel has the merit of revealing and, conversely, of integrating into the totality [...]' (Martin 1947: 31). Martin speaks also of an 'overdetermination of significations' to qualify 2020: 55 Hegel's thought of the universal 2020:76 (Martin 1947: 74).

Elliott, Gregory (1987): Althusser: The Detour of Theory, London/New York: Verso.
Foucault, Michel (1995): 'Madness, the Absence of Work.' In: Critical Inquiry 21/2, pp. 290-298. https://doi.org/10.1086/448753
Foucault, Michel (2006): History of Madness, London: Routledge.
Foucault, Michel (1995): 'Madness, the Absence of Work.' In: Critical Inquiry 21/2, pp. 290-298. https://doi.org/10.1086/448753
Gillot, Pascale (2009): Althusser et la psychanalyse, Paris: PUF. https://doi.org/10.3917/puf.gillo.2009.01
Hartmann, Nicolaï (1958 [1909]): 'Zur Methode der Philosophiegeschichte.' In: Kleinere Schriften, Band III: Vom Neukantianismu zur Ontologie, Berlin: De Gruyter, pp. 1-22.
Hegel, Georg Wilhelm Friedrich (1948): L'esprit du christianisme et son destin, translated by J. Martin, Paris: Vrin.
Heidegger, Martin (1968 [1938]): 'Ce qui fait l'être-essentiel d'un fondement ou raison.' In: Qu'est-ce que la métaphysique? Paris: Gallimard.
Heidegger, Martin (1976 [1929]): Vom Wesen des Grundes. In: Gesamtausgabe: Band 9, Frankfurt/Main: Vittorio Klostermann, pp. 123-175.
Heidegger, Martin (1977 [1927]): Sein und Zeit. In: Gesammelte Ausgabe: Band 2, Frankfurt/Main: Vittorio Klostermann.
Heidegger, Martin (1994 [1923-1924]): Einführung in die Phänomenologische Forschung. In: Gesamtausgabe: Band 17, Frankfurt/Main: Vittorio Klostermann.
Hesse, Hermann (1955): Le jeu des perles de verres, translated by J. Martin, Paris: Calmann-Lévy.
Janicaud, Dominique (2001): Heidegger en France, Vol. I, Paris: Hachette.
Lewis, William S. (2005): Louis Althusser and the Traditions of French Marxism, Oxford, Lexington Books.
Martin, Jacques (2020): L'individu chez Hegel, ed. by Jean-Baptiste Vuillerod, Lyon, ENS éditions. https://doi.org/10.4000/books.enseditions.14784
Moore, Nikki (2005): 'The Man without Work: Jacques Martin.' In: Thresholds 30, pp. 8-14. https://doi.org/10.1162/thld_a_00279
Moulier Boutang, Yann (2002): Louis Althusser, une biographie, Vol. II, Paris: Librairie Générale Française.

Wiechert, Ernst (1953): Missa Sine Nomine, translated by J. Martin, Paris: Calmann-Lévy.

Windelband, Wilhelm (1912 [1892]): 'Einleitung.' In: Lehrbuch der Geschichte der Philosophie, Tübingen: Mohr Siebeck.

'The problem itself persists': Problems as *Missing Links* between Concepts and Theories in Canguilhem's Historical Epistemology

Thomas Ebke

If one was to register the specific terminological use of the concept of the 'problem' in 20th century philosophy of the sciences and of scientific practice, it is not without a soupçon of irony that one would have to compile such an inventory. Without a doubt, or so Warren Weaver claimed in 1948, a formalised mathematics of the kind that had propelled the regime of stochastics and statistical mechanics about half a century earlier would hold the key to unlock so-called 'problems of disorganized complexity' (Weaver 1948: 538), that is to say logical situations involving a vast amount of variables, all of which display 'individually erratic' (ibid) characteristics. Conundrums of that type, Weaver argued, will turn out to be more complex to manage than 'two-variable' (ibid: 537) problems – which conveniently correspond to the binary system of the mathematics of mechanics – but it is through the recurrence of 'certain orderly and analyzable average properties' (ibid: 538) that, despite the numeric range and largely random interplay of the variables, predictions as to their standard distribution will hold true 'with increasing precision' (ibid). The reason for the relative controllability of such a system of randomly interconnected, yet homogeneous, elements is the 'disorganized complexity' of the ensemble, that is the absence of an internal *order* – an organised configuration of the components that would be irreducible to a recurring pattern of averages. As opposed to these two classes of problem constellations, 'problems of simplicity' (which can be addressed by the mathematics of mechanics) and 'problems of disorganized complexity' (mastered by stochastics), then, Weaver identifies a third type of problem that displays a strict quality of intrinsic organisation, an 'organic' (ibid: 539) immanence of the ensemble that obliterates any random conduct of its elements.[1]

Indeed, the hiatus between the logic of 'disorganized complexity' and the quality of what Weaver dubs problems of 'organized complexity' constitutes the most interesting moment within Weaver's argument: after all, the distinction between these two classes of difficulties is not based on the sheer numeric excess of the variables (which, in turn, is precisely the distinguishing mark between 'simplicity' and 'disorganized complexity'). The leap from 'disorganized' to 'organized' complexity is not tied up with an increase in the amount of empirically contingent factors that would, at one point, become impossible to prognosticate. On the contrary, the very fact that an excessive number of variables interact in empirically random constellations makes it all the more possible to single out regularities within that state of contingency and to numeralise the likelihood of recurrent events (such as, in Weaver's example of telephone communication, 'the average frequency of calls, the probability of overlapping calls of the same number, etc.', ibid). Paradoxically, it is rather the blanket extinction of empirical randomness in a system of 'organized complexity' that prevents the forecast of the conduct of its parts (and of itself, in its integrity) by means of statistical calculation. Thus, Weaver can raise the following questions that, in his account, deserve to be tackled under the label of organised complexity: 'What makes an evening primrose open when it does? [...] Why can one particular genetic strain of microorganism synthesise within its minute body certain organic compounds that another strain of the same organism cannot manufacture? Why is one chemical substance a poison when another, whose molecules have just the same atoms but assembled into a mirror-image pattern, is completely harmless? [...]' (ibid: 539). It is, of course, not insignificant at all that these textbook examples of organised complexity, which exhibit the irreducibility of 'the whole' towards the sum and the qualities of its components, stem from the world of vital phenomena – from the horizon of 'life'. In 1948, Weaver resumed Kant's hint at the peculiar teleological constitution of organisms – or

---

1 Weaver marks off this set of problems from the situation of 'disorganized complexity' with recourse to the stochastics of the billiard game: 'For example, the statistical methods would not apply if someone were to arrange the balls in a row parallel to one side rail of the table, and then start them all moving in precisely parallel paths perpendicular to the row in which they stand. Then the balls would never collide with each other nor with two of the rails, and one would not have a situation of disorganized complexity' (Weaver 1948: 538).

rather, which is, in fact, a decidedly different claim, at the way in which 'our' finite intellect inevitably judges 'organized nature' (Kant) as if it is the abode of an intrinsically teleological organisation. This specification is crucial as Weaver, abandoning the Kantian axis of reasoning, expresses his optimism that the ongoing 'advance' (ibid: 541 and passim) of the natural and technical sciences will in the end bring about empirical techniques to determine and regulate states of organised complexity. Whereas Kant had categorically severed the teleological access to the self-sufficiency of organic living beings from the mechanistic approach – which, inadequately, refers to the purposeful entirety of the organism as a totality of *partes extra partes* – Weaver clearly opts for an investment into the future progress of the natural sciences and the regime of their techniques of regulation until they will one day be able to respond to the specific intricacy of organic, that is to say emergent, forms of organisation. However, this positivistic reduction of the singular epistemic status of living phenomena (as elaborated by Kant) will not garner any systematic attention in the following analysis: this reduction is classical in its own right, and its critique can scarcely dispense with the worn-out dualisms of philosophical (or, for that matter, phenomenological) description versus empirical objectification, 'philosophical' versus 'scientific' discourse, heuristic 'openness' versus methodical 'closures' etc.

Rather, it is pertinent to take note of and to reflect on the way in which the terminology of 'problems' is tied up, at this particular juncture, with the notion of, or at least with an allusion to, the irreducible dimension of *life*. If one traces the history of the term 'problem' back to its conjuncture in 19th century philosophy and sciences, it is intriguing to learn that this particular term – as opposed, for instance, to the semantics of 'concepts' – was meant to address a delicate entanglement between the order of *objective* contents (problems as 'matters of fact') and the pole of the *subject* of scientific inquiry. To 'throw up' (see the origin of the word προβάλλειν: 'to throw something up in front of yourself') a problem amounts to irreplaceably 'having' a problem, to be, as it were, embroiled in and practically affected by the particular difficulty that poses itself. In the German tradition, Nietzsche and Simmel stand out in their emphatic pronunciation of this involvement of the subject that raises the problem *in* the problem, in defiance of the Neo-Kantian current that, during their time, identified 'problems', on the contrary, as technically specified tasks within an already established context of scientific discourse (see, for instance, the position of Richard Hönigswald, cf. Hönigswald 1931).

The summary of this specific panorama – in which problems appear on the threshold of an epistemic process that includes them in the immanence of a scientific discourse, while at the same time they remain expressive of the living subject who pursues problems, the subject to whom problems matter – sets the stage for the argument that I wish to carry out in this article. It will be my goal to show that the philosopher and historian of science Georges Canguilhem (1904-1995) elaborated a conceptualisation of what *problems* are that demarcates their role in the research process from two other elements that are also at stake in the very same process: namely, on the one hand, *concepts*, and, on the other hand, *theories*. Problems emerge and remain, as it were, on the threshold of a series of operations that bring about the specific discourse of a science, including the set of epistemic objects with which that discourse correlates. Both complementary terms, concepts and theories, need to be located within this immanence of science: it is here that they fulfill their specific functions. Yet, to the extent that concepts never fully coincide with and never entirely cover the scope of what problems are, the latter retain a quality of resistance to the conceptually restrained fabric of a science. Indeed, they point back to the *enjeu* of a vital subjectivity that initiates research processes in the first place and, in so doing, undergoes a shift of position that transforms it, the living entity, itself into an object of the sciences.

In a nutshell, the thesis around which my observations will revolve is the following: if one wishes to understand the concise status of what features as a 'problem' within Canguilhem's version of historical epistemology, one needs to explicate a triad of terms, namely the way that problems are intertwined both with 'concepts' and with 'theories' – more precisely, with scientific theories. It is only in establishing links and distinctions between these three elements and in insisting on their non-coincidence that the following argument can be upheld: according to Canguilhem, the epistemological pertinence of the problem and its fecundity within the process of the constitution of a scientific discourse harks back to the way in which a problem *outlives* or *outlasts* its conceptualisation, that is to say, the way in which it outlives the *concept*. In that sense, problems act as *missing links* between concepts and theories within the process of the formation of a science. However, Canguilhem does not at all argue that the epistemological effect of the problem is to provide the inaugural piece in the genesis of a scientific discourse that could in the end be considered as self-sufficing, immanent and complete. Rather, it is Canguilhem's very distinction of problems as opposed to concepts *within* the dynamism of conceptualisation proper that enables and motivates him to mount a critique of scientism precisely by insisting on the *historicity* of the generation of scientific discourses. It is against the background of these distinctions that Pierre Macherey echoed Canguilhem's famous definition of philosophy as the 'science of solved problems', *la science des problèmes résolus*, as Canguilhem phrased it in his study on the formation of the concept of the reflex in the 17th and 18th centuries (Canguilhem 1955). On Macherey's reading, the 'philosophie du concept' pursued by Canguilhem constitutes '*la science des* problèmes indépendamment de leur solution' (Macherey 2008: 56). This expression, after all, hits an insight that seems to be deeply characteristic of Canguilhem's historical epistemology in its entirety, that is the insight that *philosophy* addresses an element – namely problems – which is elaborated, as it were, from the inside and within the boundaries of a science, yet without reproducing and sharing the very means that science summons up in coming to grips with that certain problem: those means being the concepts of science. It seems helpful to reiterate the gist of these observations: although Canguilhem locates philosophy *immanently*, at a point that finds itself in the midst of the very operations of scientific rationalisation, he does so only to explicate a sharp split between philosophy and science. Philosophy, for Canguilhem, presents itself as 'a science of problems' in such a way that the problems (re-) appear in their independence from the solutions that scientific discourses have endowed them with. Indeed, it is crucial to underscore the verb *reappear* at this juncture and to draw attention to its temporality: to return to the problems that had been at stake at the outset of a process of scientific conceptualisation is in itself a historical procedure whose effect it is to reinstate what has (already) been 'framed', that is to say operated on and thus rationalised by means of scientific concepts. One would not be entirely misled if one concluded that Canguilhem restores on the part of the objects of science what, in an inevitable reduction, had been taken (that is: *abstracted*) from them in the very process that turned them into epistemic objects in the first place – namely their problematic status, their quality as problems according to which they remain specifically external to the concepts that relate to them and express them under the conditions of science. What Canguilhem dubs a 'historical epistemology' is thus a decidedly philosophical undertaking while, at the same time, his conception of philosophy takes on the irreducible form of an epistemology that cannot but proceed historically.

In what follows, I will attempt to elaborate the point that historical epistemology, in Canguilhem's definition, is a type of reflection that situates itself at the crossroads of the history of science and a philosophy of values: in a certain discord with the wording used by Macherey, I will suggest thinking of this epistemology not as the *science* of 'problems independent of their solutions', but as a reciprocally historical and philosophical gesture that separates the *problems* of scientific rationalisation from their *conceptual* and *theoretical* solutions, which is precisely what the sciences engender. In an inaugural step, however, it will be crucial to expound the specific genealogy and meaning the term 'problem' takes on in the writings of Canguilhem.

The epistemological dispositive of Canguilhem's problematology: the fissure between *protasis* and *problem*

In his biography on Michel Foucault, Didier Eribon portrayed Georges Canguilhem as a clandestine intellectual who, although constantly looming but in the *background* of a scenery that boasted more flamboyant protagonists (such as Foucault himself), obliquely shaped the entire agenda of philosophical discussions in France in the 1950s and 1960s (Eribon 1989: 232). According to Eribon, the key antagonism that remained at work underneath the major controversies of these decades was the split between, on the one hand, Canguilhem, and on the other, Jean-Paul Sartre. Thirty years after Eribon evoked that picture, at a time when Canguilhem was still alive, it is fair to say that this philosopher has been elicited from the relative obscurity that surrounded him *à l'époque*. The systematic reception of Canguilhem's thought can be rather neatly traced back to 1996, which witnessed the publication of Gilles Renard's monograph on Canguilhem's rendition of historical epistemology (Renard 1996). In the aftermath of this rediscovery, one axis of reception that has been particularly prominent is a line of research that inscribes the project of Canguilhem into a 'Bergsonian' heritage (see Osborne 2003, During 2004, Worms 2009, Schmidgen 2014, Delitz 2015). According to this reception, Canguilhem's notions of the normativity of life and of the primacy of practice within scientific inquiry are strongly tied to the modern vitalism of Henri Bergson, who stands aloof in the landscape of 20th century French philosophy due to his singular insistence on a philosophy of life.

The emphasis on Canguilhem's 'Bergsonian' filiation has exerted an interesting impact on contemporary readings of Canguilhem. By and large, Canguilhem's specific tackling of the question of what problems actually are has been reduced to the Bergsonian claim that 'genuine problems demand the creation of the concepts that will be used to posit them' (Bowden 2018: 48). On that reading – which certainly has its merits – the decisive point is the practical quality and efficacy of problems, which, rather than being the mere correlates of scientific practice, essentially engender the techniques, including the conceptualisations, that science requires in order to (literally) come to terms with 'its' problems (ibid). But it would be a cliché to associate Canguilhem's understanding of problems too tightly with this 'Bergsonian' trajectory, which cannot but amount to the idea of a radical primacy of life over and against its objectifications via science. Instead, it can be helpful to note that what seems to linger in Canguilhem's vocabulary of 'the problem' is a decidedly *Aristotelian* echo: in view of the logical issue of problems in his *Topics*, one can identify at the centre of this tract the motion of dialectical reasoning (see Margel 1997: 160-162). Aristotle, in fact, makes a distinction between πρόβλημα and πρότασις: the first term, the problem, represents that *upon which* (τά περι ων) a dialectical train of arguments is grounded: that which is thrown up in the logical form of disjunction. The second term, by contrast, refers to that element *out of which* (τά ἐξ ὡν) he who draws the logical conclusion can base the dialectics (see ibid for this entire reconstruction): that is, in the sense of the classical syllogism, the premises. In other words: the protasis is the active heuristic operation that takes root in what, in logical priority, had been thrown up before *the one who constructs the dialectical argument*. That is to say, technically speaking, the premises of a dialectical construction take on the *form* of the premise only by means of a (protatic) reply to a (problematic) question, that has, in logical antecedence, been raised and posed, thereby eliciting the dialectical motion of arguments.

At first glance, it might be difficult to discern the point of interest that is at stake here: if a fissure constantly remains between the problem that is brought up – or rather, 'projected' (ibid: 170) – in a proposition and the *protasis* which takes up that very problem under the form of syllogistic premises, then this transition crucially implies a discursive 'space of controversy' (ibid: 169: 'espace d'une controverse') in which interlocutors confront each other as adversaries. In a first step, this internally polemical or controversial structure explicates itself to the extent that a problem does not simply coincide with any enunciation whatsoever; rather, it is precisely an enunciation constituted in a form that is open to discussion: 'Ought one rather to obey one's parents or the laws, if they disagree?' (Aristoteles 1984: 176). In other words, the problem suspends any reference to facticity by turning a presumed fact into a moot subject, a question that elicits logical reasoning on behalf of the interlocutors. On a second level, this enunciation that has shifted into the status of a problem invites a multiplicity of heterogeneous (rational, conceptual) solutions, including the acknowledgment (or the denial) of the *problematic status of that which has been thrown/brought up* – that is to say, of the πρόβλημα itself. The fissure between and the passage from problems to premises, then, is intrinsically polemical, it is permanently open to discussion. Yet, it would be justified to speak of a genuine Aristotelian 'dialectics of problems' (cf. Marge 1997), because what unites the problem with the premise(s) and what in fact transforms the problem into the premise through the operation of πρότασις is the logical form of interrogation. By posing the problem in the form of a question to the interlocutor binds the one who asks and the one that is called upon to answer to the discursive standard of the *judgment*. It is the interpellative function of interrogation that guarantees the formal connection of the problem and the premise, the projection and the proposition (see ibid: 173-174).

This recourse to Aristotle's *Topics* might in fact be conducive to an adequate reading of the systematic role played by the term 'problem' within Georges Canguilhem's historical epistemology. At least *one* impact that I hope my observations in this paper may induce rests in the claim that, in opposition, or rather as an amendment to, the Bergsonian interpretation of Canguilhem's 'philosophy of the problem', one should elaborate on the intrinsically epistemological dimension, that is to say, on the Aristotelian vein of Canguilhem's interest in *problems*. According to Aristotle, the eminent logical function of problems hinges upon the 'dialectical' quality of their logical solution, which transposes the discussion of the problem into the twofold form of interrogation and, as its correlate, judgment. In the secondary literature on Canguilhem, which, in some sort of pragmatist reflex, too frequently focuses on the vital(-ist) foundation of the semantics of problems, anchoring it in Bergson and/or Bachelard as Canguilhem's major sources, this epistemological dispositive tends to be underestimated (see During 2003, Osborne 2003, Schmidgen 2014, Feldman 2016, Bowden 2018). Yet, to overlook this dispositive would mean to fall short of the argument over why Canguilhem's project of historical epistemology is an intrinsically political and in itself a normative intervention. Before this position can be spelt out in more detail, however, it is indispensable to go back to at least two of the most prominent quotations from the writings of Georges Canguilhem that foreground the notion of the 'problem.'

Theories, concepts, problems

In one of the most famous wordings from his magisterial study *The Normal and the Pathological* (1943), a passage that stems from the original introduction to the book, Canguilhem underscores that the reason why he studied medicine in the first place and later earned a doctorate in that discipline was his expectation that medicine might 'provide precisely an introduction to concrete human problems' (Canguilhem 1978: 6). This quotation, as well as the one that follows, lends credence, all in all, to what one might apostrophise as the *vitalist* key to reading Canguilhem: on this view, problems designate an objectivity that is dealt with by a scientific discourse, yet in that grasp outlasts the very rationality of such a discourse, which cannot but be based upon an interplay of concepts and a theory that guarantees their coherence. A problem seems to represent a difficulty of an intrinsically *technical* register – an underlying issue that propels the immanent 'solutions' generated by a science, while at the same time outwearing any such termination. It is, of course, not by accident that this distinction is evoked by Canguilhem in the direct context of a reflection on the normativity of the living organism as a factor that thwarts its full objective explication within a scientific physiology. More concretely, at one moment in his book Canguilhem tackles the question of whether Claude Bernard expressly intended, in the formulation of his own physiology, to blur the strong, qualitative idea that the pathological states of the organism are in themselves genuinely normative states that are irreducible to the states and conditions of the 'healthy' organism. Tending to credit Bernard with exactly such a strategy, but at the same time with a certain hesitation as to the legitimacy of such a suspension, Canguilhem continues with the following interesting remark: 'This ambiguity is certainly instructive in that it reveals that the problem itself persists at the heart of the solution presumably given to it. And the problem is the following: Is the concept of disease a concept of an objective reality accessible to quantitative scientific knowledge? Is the difference in value, which the living being establishes between his normal life and his pathological life, an illusory appearance which the scientist has the legitimate obligation to deny?' (ibid: 36).

Problems, Canguilhem contends here, 'persist at the heart of the solution' they have been endued with; that is to say, a problem outlasts the conceptual operation which does not only raise the problem but, in so doing, renders it intelligible. However, it would be mistaken to envisage this deferral of concepts vis-à-vis the problem as a purely derivative relationship (as if the concepts of a science cannot but fall short of their underlying problem): instead, it is equally important to recognise that problems, rather than existing aloof from scientific rationality, always appear on the threshold of a science, permanently on their way to a science's immanent nexus of a theory and its concepts. Now, a close examination of the secondary literature on Canguilhem's problematology (see sources above) demonstrates that most readings of his disjunction between problems and concepts stop at the 'vitalist' conclusion and its twofold logic. At any rate, the vitalist reading of the way Canguilhem winnows problems from concepts does not only accentuate the role of the problem as a technical obstacle within the discourse of science, but also identifies *life* as the epitome of any productive (viz. normative) force that defies all positivisation. Thus, by definition, life features *par excellence* as the intrinsically generative process of the 'formation of forms' (Canguilhem 2008: XIX): Whereas the analytic determinations brought about by the discourse of science cannot hold good but for the terminal forms of vital processes, life comes into view as the formative process itself from which all those forms originate. Therefore, the primary mode in which the specificity of life as an intrinsically normative phenomenon becomes expressive is the mode of *techniques* (ibid).

It will be my core observation in this paper, however, that in terms of a corrective or, as it were, an amendment to this vitalist stance, one ought to reevaluate the Aristotelian legacy in Canguilhem's appropriation of the semantics of the 'problem'. This legacy provides Canguilhem's approach to the sciences with a fully fledged *political epistemology* that revolves precisely around the fissure between πρόβλημα and πρότασις: the latter term representing an active reply to an antecedent *problematic* question that can or cannot subsequently be carried out in the logical form of the premise. Xavier Roth (Roth 2013) has recently drawn attention to the lineage in Canguilhem's thinking that connects his approach with the (French) 'style de pensée réflexif' (ibid: 129). Among the most renowned protagonists of this intellectual current, Jules Lachelier (1832-1918), Léon Brunschvicg (1869-1944) and Émile-Auguste Chartier aka 'Alain' (1868-1951) stand out, and Roth gives a helpfully concise idea of what is central to the philosophical tradition of the *analyse réflexive*: this tradition essentially represents an epistemology which insists on the irreducibility of values to facts and on the perpetual challenge to the human spirit to elude the reign of facts by the very acts of judging their genesis and validity (ibid: 130-131). 'Reflexive analysis' thus designates the bending back (*re-flexio*) of the fact to its intrinsic axiology, to the judgmental, that is to say evaluative operations that are sedimented in 'matters of fact'. Nowhere has Canguilhem spelt out the methodology of his 'reflexive analysis' of the sciences more clearly than in his early *Traité de Logique et de Morale*, co-written with his colleague Camille Planet in 1939, when both authors were employed as teachers of philosophy at a *lycée* in Toulouse. It is in this treatise that Canguilhem gives the following portrait of science as a rationale of substituting practical obstacles with (temporary) inconsistencies on the level of discursive conceptualisation: 'Toute science est analyse des *obstacles* que l'existence supposée d'une Nature dresse dans l'expérience devant nos désirs; pour le jugement théorique qui a ainsi *décidé* de se constituer, l'échec prend la form examinée plus haut, celle de l'erreur, en se considérant uniquement comme affirmation de réalité reniée par l'objet même.' (Canguilhem/Planet 2011: 653)

The quintessence of this wording is sufficiently explicit: the discourse of science hinges upon an epistemological *decision* to confront practical impediments as immanent issues within a process of conceptual rationalisation. What is more, it is not by accident that Canguilhem takes up the role of *judgments* in this context: on a first level, judgments are located within and effectuated by the scientific discourses themselves. It is in the interest of the consummation of the scientific practice to treat technical obstacles, bluntly speaking, as immanent moments of the working of the concept. Yet, on a second level, there is also a historical judgment at work here, that is to say the evaluation of the historian of science who judges the judgments of the scientists themselves. It is this twofold gesture which, as far as I can see, structures Canguilhem's idea of 'reflexive analysis.' Another quotation from the *Traité* corroborates Canguilhem's 'pre-vitalistic', reflective starting point of a genuine philosophy of the problem: '[Q]uand nous commençons de dissocier le chaos perceptif en y cherchant des "genres" de choses, nous sommes amenés, pour comprendre les choses les unes par les autres, à multiplier non seulement ces genres, mais les "points de vue" sous lesquels ils nous *apparaissent* et nous constituons ainsi des *concepts*. [...] Les concepts pris tels quels ne sauraient être des vérités; au même titre que les sensations, et bien que formés d'autre manière, ils proposent des *problèmes* à qui essaie des les comprendre et pour cela des les ordonner.' (ibid: 721)

This is a particularly rich quotation that needs to be unpacked with some caution, not least because it eventually helps to elucidate the division between *problems*, *concepts* and *theories*. To begin with, on inspection of this quote, it seems that two different ways to understand what a 'concept' does or is can be ruled out on its basis (viz. Schmidgen 2014). On the one hand, concepts do not represent reality; they have no semantic and no referential function. But on the other hand, neither can they be adequately interpreted as 'constructions' (ibid: 238), not even in Kant's gentler, decidedly non-constructivist sense of a priori determinations of thought that are, in the last instance, anchored in the transcendental apperception of consciousness. Rather, as both Canguilhem's and Planet's choice of words and their reading by Henning Schmidgen suggest, concepts normatively *make* something *visible*, they 'produce realities and perceptions, and stimulate activity' (ibid). A concept seems to inscribe into the phenomenon it addresses a distinct 'point of view', a distinctive operation that imposes a normative decision on the one who approaches the phenomenon via this concept, such as a researcher in an empirical situation of a scientific practice. As a textbook example, one might think of the concept of the reflex and the way that it necessarily involves an elementary opposition, namely the opposition between voluntary and involuntary movements. After all, the productive performance of the concept consists of the way it raises or proposes a problem – to take up Canguilhem's and Planet's expression here. Importantly, the two young authors continue to argue, on the following page of their treatise, that concepts, in this light, indeed already imply 'connaissance' (Canguilhem/Planet 2015: 722) – which I hesitate to translate here with 'knowledge' – but precisely not 'connaissance *vraie* ou scientifique' (ibid). After all, then, one seems to be justified in pointing out that concepts are on the road to more precise knowledge. Yet, what keeps a concept aloof from 'connaissance' in the strict sense is that it does not, in and by itself, generate an internally coherent system of judgments that explicitly order and classify the phenomena under the perspective that this particular concept is able to open up. In other words concepts supply a sketch of the solution which they tend to give to the specific problem that correlates with them. But here comes what Canguilhem and Planet add to their portrait of what, in their book from 1939, a concept is: 'C'est que, de la subjectivité impliquée dans l'expérience originelle, les concepts gardent une sorte de contingence, et même une instabilité: tel pense inoffensif ce que tel autre juge dangereux, durable ce qui celui-ci croit précaire, etc.' (Canguilhem/Planet 2015: 794)

This is indeed a salient point in Canguilhem's entire conception of epistemology: on a level that is not already included in the immanence of the thorematic operations of a science, concepts continue to implicate a *contingency*, or, as the quotation has it, a precarious openness – and this contingency is precisely the share of the *problem*. No ultimate determination of the problem that is at stake in connection with a specific concept can ever be reached, and while concepts seem to invite and elicit research that is conducted under the genuine conditions of science – that is, within a discourse that aims at producing or speaking 'the truth' – they are never fully reducible to the realm of science. At this juncture, it is no longer complicated to discern the disjunctive operation that culminates in the slip between problems and concepts. The reason for this fissure is the internal tension between πρόβλημα and πρότασις: the transition from problems to premises always rests upon a normative act (the πρότασις), an implicit act of judgement that decides to accommodate the problem inside a conceptual regime. And it is precisely the flip side of this operation that Canguilhem (alongside Planet) calls to mind with his usage of the term 'problem': to remit the discourse of science to its *problems* as opposed to its concepts amounts to unveiling the normative judgments, and the historicity of these judgments, that structure the discourse of science itself. In other words, the recourse to the problem is the key to that 'reflexive analysis' which explicates the epistemic decisions of a science and, in so doing, itself judges those decisions as historical *enjeux*.

In his paper on the topic, Henning Schmidgen contends that whenever Canguilhem speaks of a 'concept' in the technical sense, there is always an interplay between a phenomenon, a denomination and a definition at work (Schmidgen 2014: 246). For example, while we encounter the word 'reflex' in the sense of a *denomination* in the writings of Descartes, it is not the case that this word carries with it a fully fledged *definition* of the object it addresses. In fact, as Canguilhem shows in his study about the formation of the concept of the reflex in the 17th and 18th centuries (Canguilhem 1955), Descartes did not conceive of a homogeneity between sensory stimulation and motor reactions, whereas in the latter half of the 17th century, notably in the writings of the British physician Thomas Willis, it is exactly the functional symmetry of these two cycles that is conceptualised, that is to say determined as a reflection, a reflux, or an echo, by means of the word 'reflex'. Only then does the word 'reflex' take on the terminological function of a 'concept': it begins to address the phenomena of voluntary and involuntary motions under a qualitative distinction between these two groups of phenomena. In other words, it classifies and stabilises experience through the lens of this distinction.

Pierre Macherey has drawn rich conclusions from this methodology in Canguilhem's writings. The following quotation from his above-mentioned piece on Canguilhem's philosophy of science speaks for itself: 'Un *concept*, c'est un mot plus sa définition; le concept a une histoire; à un moment de cette histoire, on dit qu'il est formé: quand il permet d'instituer un protocole d'observation' (Macherey 2009: 59). One can retain the idea that concepts have a history only to the extent that one speaks of something that might have been, and can always be, conceptualised in ways and terms different from the path that a science historically took as a matter of fact. This factor is precisely what Canguilhem has termed, on more than one occasion, a *problem*. However, one needs to be concise on this point: the notion of the problem is not part of a position that one might describe as an epistemic realism. Problems are no realities that would exist outside of and preceding the work of conceptualisation. In this regard, Canguilhem is very outspoken and he sides with Duhem, Rey and Bachelard in saying: 'Le fait n'est pas ce dont la science est faite, mais ce que fait la science en se faisant' (Canguilhem 2015: 371). Thus, the difficulty in coming to terms with the role of *problems* as opposed to *concepts* in Canguilhem's epistemology resides in the fact that Canguilhem does indeed separate these two poles, but that does not mean that he grants the problem an ontological status anterior to the process of conceptualisation. Strictly speaking, then, problems are not phenomena. Rather, a problem is a specific task that a conceptual operation tosses up in front of itself as something which requires a solution – which is supposed to need a solution that is not yet, or may be no longer, in place. All of this is particularly relevant under the aspect of time, in the perspective of temporality: although Canguilhem aims at explicating *that* dimension which presents itself, after all, as exterior and irreducible to science, he does not at all pursue the direction of a phenomenology of lived experience, and to some extent he does not even (strictly) associate this dimension with his philosophical vitalism. What one needs to understand, then, is the immanent critique of scientific rationalities, a project that, by the way, does not exclude, but rather implies the affirmation of the peculiar normativity of the sciences. In the last step of my argument, I will briefly try to betoken the fabric of Canguilhem's historical epistemology in its 'pre-vitalist' configuration that is not yet, at least not explicitly, pervaded by the argument of biological normativity, but rather oriented towards the 'reflexive analysis' of the *problems* of science.

The upshot of the reflexive analysis: moments of stability, moments of crisis

Pierre Macherey has reminded us of something pertinent, namely of Canguilhem's claim that concepts are actually *born* (Macherey 2009: 58). Therefore, what needs to be reconstructed are the constellations and the normative choices, the *enjeux* which have a hand in the birth of a concept. The temporality of the concept does not coincide with the history of a scientific discourse or of any scientific discourse which operationalises that concept. In order to fully account for this non-coincidence, one would have to earmark three breakages which a historical epistemology in the lines of Canguilhem deals with and draws our attention to. The first breakage is the irreducibility of problems to concepts: concepts tackle problems through an act of πρότασις, thus transforming them from technical obstacles into the immanent subject matter of scientific rationalisation. It is on this level that Macherey can introduce the idea of philosophy as a questioning of problems in their independence from their solutions. The second breakage concerns the way that scientific discourses implicate problems in a theorematic regime. In this perspective, Canguilhem insists, as Macherey has justly shown, not only that a theory has its moments of crisis, but actually even a *birth*, the birth of the theory being the decision of science to tackle problems conceptually, driven, as it were, by something that one might call a *volonté de la vérité*, a will to produce and to speak truth. Thirdly, however, there is another breakage, which, too, has to be reconstructed historically, and that is the point where science spills over to arenas that are outside of science per se, for instance, to the field of techniques as opposed to science. The example of the rise of positivist physiology, which has eclipsed the genuinely technical and practical dimension of medicine, is particularly telling in this respect.

And one finally begins to understand, too, that this philosophy is intrinsically tied to and can only constitute itself as a history of science. Philosophy raises and highlights problems in a gesture that is not driven by the will to solve them, by the *volonté de la vérité* that is immanently constitutive of science. On the contrary, philosophy solicits a precise historical reconstruction of the distinct ruptures and normative stakes in the process of formation of scientific knowledge. To understand what is at stake here, a glimpse at the philosophical articulations of 'Canguilhem avant Canguilhem' (J.-F. Braunstein) can be particularly precious: it is here that the epistemology of reflexive analysis is foregrounded, an operation that lingers on the threshold between problems and concepts. Simultaneously, this analysis exposes and normatively questions the judgments, that is the valuations, that drive a science to pursue problems within the regime of concepts: if it is true that 'the problem itself persists' at the heart of its scientific solutions, the task of historical epistemology can only reside in the constant liquefaction of facts with a view to the values that remain sedimented inside them. This dynamic operation, in turn, is rendered possible by the immanent split of the process of problematisation, the split that separates the πρόβλημα from its πρότασις. Only four years after the *Traité*, co-written with Camille Planet, Canguilhem will move to an integration of his reflexive analysis into a philosophical vocabulary of life and of biological normativity. But the oblique model of and complement to this vitalism, it seems to me, needs to be identified in his epistemology of 'reflexive analysis'. It is only along the lines of such an analysis that the genuinely vital dialectics of the sciences, their interplay between 'moments of crisis' and 'moments of stability', can garner their full credit.[2]

---

2 'Du moment que nous parlons de science, c'est-à-dire de connaissance vraie, nous sommes dans l'ordre de cette opération mentale qui seule peut être dite vraie ou fausse, c'est-à-dire l'établissement dans le jugement d'un rapport entre concepts. [...] Le fait n'est pas ce dont la science est faite, mais ce que fait la science en se faisant. [...] Donc s'il est vrai qu'il n'y a de science que sous forme de théorie, c'est-à-dire de démonstration, et qu'il n'y a pas de démonstration sans principes, nous dirons que la science expérimentale est celle qui va à la recherche de ses principes. Le concept de science expérimentale est un concept mixte qui retient à la fois la relation de la science à l'expérience comme au problème à résoudre et la relation de l'expérience à la science comme au théoreme, c'est-à-dire au problème résolu. [...] Mais si toute science est théorie, composition rationnelle, c'est-à-dire déductive de

References

Aristotle (1984): 'Topics.' In: The Complete Works of Aristotle, Vol. 1, Princeton: Princeton University Press, pp. 167-276.
Bowden, Sean (2018): 'An anti-positivist conception of problems: Deleuze, Bergson, and the French epistemological tradition.' In: Angelaki 23/2, pp. 45-63. https://doi.org/10.1080/0969725X.2018.1451461
Canguilhem, Georges (1955): La Formation du concept de réflexe aux XVII siècle et XVIII siècles, Paris: PUF.
Canguilhem, Georges (1978): On the Normal and the Pathological, Dordrecht/Boston/London: Reidel Publishing Company.
Canguilhem, Georges (2008): 'Introduction: Thought and the Living.' In: The Knowledge of Life, New York: Fordham University Press, pp. XVII-XX.
Canguilhem, Georges (2015 [1948]): 'Rôle de l'histoire des sciences dans la philosophie des sciences: l'établissement des faits fondamentaux de la dynamique.' In: Œuvres complètes, Tome IV: Résistance, philosophie biologique et histoire des sciences 1940-1965, Paris: Vrin, pp. 369-391.
Canguilhem, Georges/Planet, Camille (2011 [1939]): 'Traité de logique et de morale.' In: Georges Canguilhem: Œuvres complètes, Tome I: Écrits philosophiques et politiques 1926-1939, Paris: Vrin, pp. 633-924.
Delitz, Heike (2015): Bergson-Effekte: Aversionen und Attraktionen im französischen soziologischen Denken, Weilerswist: Velbrück. https://doi.org/10.5771/9783845277523
During, Elie (2004): 'A history of problems: Bergson and the French epistemological tradition.' In: Journal of the British Society of Phenomenology 35/1, pp. 4-23. https://doi.org/10.1080/00071773.2004.11007419
Eribon, Didier (1989): Michel Foucault, Paris: Flammarion.
Feldman, Alex (2016): 'The concept in life and the life of the concept: Canguilhem's final reckoning with Bergson.' In: Journal of French and Francophone Philosophy – Revue de la philosophie française et de langue française XXIV/2, pp. 154-175. https://doi.org/10.5195/JFFP.2016.775
Hönigswald, Richard (1931): Grundfragen der Erkenntnistheorie: Kritisches und Systematisches, Tübingen: Mohr Siebeck.

---

concepts, il y a dans la constitution des théories des moments de stabilité et des moments de crise. Il y a une naissance, une vie et une mort des théories.' (Canguilhem 2015: 370-371)

Macherey, Pierre (2009): 'La philosophie de la science de Georges Canguilhem: Épistémologie et histoire des sciences.' In: De Canguilhem à Foucault: La force des normes, Paris: Éditions La Fabrique, pp. 33-70.

Margel, Serge (1997): 'Problématique et paradoxe: La définition aristotélicienne du problème.' In: Revue de Philosophie Ancienne 15/2, pp. 159-188.

Osborne, Thomas (2003): 'What is a problem?' History of the Human Sciences 16/4, pp. 1-17. https://doi.org/10.1177/0952695103164001

Renard, Gilles (1996): L'épistémologie chez Georges Canguilhem, Paris: Nathan. https://doi.org/10.3917/nath.renar.1996.01

Roth, Xavier (2013): Georges Canguilhem et l'unité de l'expérience. Juger et agir, 1926-1939, Paris: PUF.

Schmidgen, Henning (2014): 'The life of concepts: Georges Canguilhem and the history of science.' In: History and Philosophy of the Life Sciences 36/2, pp. 232-253. https://doi.org/10.1007/s40656-014-0030-1

Weaver, Warren (1948): 'Science and Complexity.' In: American Scientist 36, pp. 536-544.

Worms, Frédéric (2009): La philosophie en France au XX siècle: Moments, Paris: Gallimard.

Compositional Methodology: On the Individuation of a Problematic of the Contemporary

Celia Lury

Introduction

The primary focus of Gilbert Simondon's writings is the problematic of individuation (1992; 2017); a subsidiary but integral dimension of this focus is a concern with *the individuation of a problematic*. In this essay, I explore some of the methodological aspects of this process. At a fundamental level, approaching the construction of a problematic as a process of individuation, Simondon provides an alternative to theories of knowledge in which the possibility of knowledge is grounded in the constituting activity of the knowing subject. As he puts it, 'We cannot *know individuation* in the common sense of the phrase; we can only individuate, individuate ourselves and in ourselves' (1992: 317).

In Simondon's approach to the individuation of a problematic, metaphysics and logic are merged in what is called transduction, that is, a recursive analogical operation in which the process of individuation 'between the real exterior and the subject is grasped *by* the subject due to the analogical individuation of knowledge *in* the subject' (Simondon quoted in Combes, 2012: 9; my italics). Transduction is the analogical and self-grounding dimension of the procedure of thought. As Adrian Mackenzie explains,

> Every transduction is an individuation in process. It is a way something comes to be. Importantly, transduction refers not only to a process that occurs in physical, biological or technical ensembles as they individuate. It also occurs in and as thought. Thinking can be understood as an individuation of a thinking subject; not just something that someone who thinks does. (2002: 18)

The individuation of a problematic is always double, both ontogenetic and epistemological. As Combes put it in her influential account, for Simondon, 'thought is nothing more than one of the phases of being-becoming, because the operation of individuation does not admit of an already constituted observer.' (2012: 7)

In what follows my concern is to explore the implications of the onto-epistemological doubling by focusing on issues of methodology. Given this concern, I risk falling into a kind of technocratic conception of the problematic, of inhabiting the position of a functionary (Flusser 2014), but I hope to avoid doing so by following Simondon in refusing to understand methodology as a principle or a set of principles that can simply be applied or put into effect. Instead, the aim is to see methodology as a principle – or perhaps better – and, to adopt Simondon's terminology, an operation, that *is itself constituted as it happens*: 'the transposition of the scheme is in turn accompanied by a composition of it' (Simondon, quoted in Combes 2012: 13). I put forward the term 'compositional methodology' to describe the dynamic and methodologically constitutive dimensions of the individuation of a problematic (Lury, forthcoming), the twisting of process into practice.

The use of the term 'compositional' is intended to draw attention to Simondon's distinctive notion of form. His criticism of hylomorphism – in which a pre-existing form is imposed on matter from outside – is well known. In Simondon's view, such an understanding is inadequate in that it does not recognise the potentials that are always emerging in a process of individuation. To acknowledge this potential, he proposes a concept of form related to the activity of in-formation. In this understanding, information designates 'the very operation of taking on form, the irreversible direction in which individuation operates' (Combes 2012: 5). This understanding of form as an operation – a continuous, variable process – is also present in other process thinkers, including A.N. Whitehead, who says that the comprehension of process requires 'an analysis of the interweaving of data, form, transition, and issue' (Whitehead 1968: 34).

A problematic of the contemporary

My aim here is to consider the methodological aspects of the individuation of a problematic by addressing the individuation of a problematic *of the contemporary*. Peter Osborne (1995) provides a note of caution for such a project when he suggests that the space of the contemporary is constituted as an illusory present in much social and cultural theory. He gives as an example those analyses that make use of the term 'new media' as if 'new-ness' provided its own context. This present-ism, he suggests, is a consequence of theory's modernism and its negation of the past. Acknowledging the pitfalls of such an approach, I propose instead to elaborate what might be involved in the individuation of a problematic of the contemporary by drawing on Paul Rabinow's anthropology of reason, which explicitly aims to move beyond modernity as a metric of inquiry.[1] Rabinow presents the contemporary as an assemblage of old and new elements and their interactions and interfaces: 'The contemporary is a moving ratio of modernity, moving through the recent past and near future in a (nonlinear) space that gauges modernity as an ethos already becoming historical.' (2009: 2)

In this way of thinking, the individuation of a problematic of the contemporary is emergent, where emergence refers to 'a state in which multiple elements combine to produce an assemblage, whose significance cannot be reduced to prior elements and relations' (2009: 2). Simondon himself uses the term contemporary to describe the individuation of living individuals, in contradistinction to that of physical individuals: he says that the living individual is contemporary with itself whereas a physical individual contains a past that is 'radically "past"' (1992). The restriction of this understanding of the contemporary to living entities is not adopted here since, methodologically, it is important to recognise the heterogeneity of milieu in which the individuation of a problematic takes place and acknowledge the significance of modes of conscious and non-conscious cognition that cut across the distinction between the living and the non-living (Hayles 2017). Although some scholars (Back and Puwar 2012) have proposed the term 'live methods' to galvanise methods across the social sciences, what is proposed here is a little different (though not incompatible): that is, a compositional methodology is

---

1 Relatedly, Rabinow says, 'it is only through discovering and giving form to elements that are already present that the inquiry can proceed' (2009: 9).

proposed in which the individuation of a problematic is a processual ensemble of living *and* non-living entities.

Having laid out some general methodological characteristics of the individuation of a problematic of the contemporary, let me now briefly introduce the notion of epistemic infrastructures to orient the discussion of the environment in which the individuation of problematics takes place. The term 'infrastructure' highlights the ways in which individuation requires and installs 'material supports' in the world, including 'buildings, bureaucracies, standards, forms, technologies, funding flows, affective orientations, and power relations' (Murphy 2017: 6). The term epistemic is used to signal that what is at issue is the nature of knowledge, justification and belief rather than (pre-formed) knowledge as such. Indeed, the relation of knowledge to truth, and the complex relations between knowledge, faith, feeling and belief are central to discussions of the individuation of the problematic of the contemporary (for different perspectives on these issues see Simone 1994; Connelly 1999; Esposito 2013; Blencowe 2015).

Crucially, infrastructures are never self-contained or discrete: they themselves leave legacies that impinge on environments, atmospheres and other material sites through complex interrelationships of energy transference, waste disposal and economics (for example, data centers globally currently account for 2% of global greenhouse emissions equivalent to aviation)' (Corby 2017: 368). And of course they are always changing. Consider, for example, the changes described by Alberto Corsín Jiménez (2014) in his discussion of beta or open source urbanism. By this he means the ways in which citizens are 'wiring the landscape of their communities with the devices, networks, or architectures that they deem worthy of local attention or concern' (2014: 342). He situates these developments in relation to the 'new economy of open knowledge' emerging from organisational forms, such as peer-to-peer networks of collaboration, and in doing so identifies the possibility of *a right to infrastructure*.

Corsín Jiménez identifies three dimensions to this right. One, conceptual: projects in open source urbanism populate urban ecologies with digital and material entities whose emergence destabilises classical regulatory distinctions between public, private or commercial property forms, technologies and spaces. Two, technical: open source urban projects are built on networks of expertise and skills that traverse localised boundaries. Three, political: open source projects transform the stakes in modes of urban governance. In an open source project, a community may assume political and expert management of its infrastructure. By bringing these three dimensions together, he suggests, it becomes possible to read the right to infrastructure as a verb, not a noun:

> The process of *infrastructuring* makes visible and legible the languages, media, inscriptions, artefacts, devices, and relations – the betagrams – through which political and social agencies are endowed with [...] expressive capacity. (2014: 357)

But, as Corsín Jiménez also acknowledges, both the right to and the capacity for infrastructure are unevenly distributed. There is a precarious politics associated with the articulation of this right and the associated redistribution of expertise. For example, in the UK and more widely in Europe, academics in all disciplines are now routinely encouraged to extend collaboration outside the academy, not simply to engage representatives of business, government and the third sector, social movements and the public, but to invite them to participate in research activity as co-producers of knowledge. On the one hand, it is part of a shift in emphasis from the experimental as a knowledge-site to the experimental as a social process. On the other hand, it is also an unequal playing field, in which the nature and characteristics of the social are being redefined (Marres, Guggenheim and Wilkie 2018). At the same time, 'users' (that is, most of us who engage with digital media as part of our everyday lives) are increasingly required to take part in/be part of a variety of genres of participation, including tests, trials, games, competitions, experiments, quizzes and so on, also including various forms of tracking and tracing, where we are not so much participating as being participated. In these practices, knowledge-making is implicitly and explicitly tied to the creation of epistemic cultures of increasingly diverse, distributed heterogeneous kinds (Knorr Cetina 1999), including (multi-sided) markets. As the traditional entry and exit points of knowledge-creation become less clear 'information' asymmetries proliferate.

The politics of infrastructuralism is also at issue in the emergence of what are called global challenges or global problems. In these uses, the global is sometimes understood as size or magnitude, that is, some problems – such as climate change or disease – are understood to be so 'big' as to be global. In addition, the term is sometimes extended to a concern with the heterogeneity of the actors said to be required to address such problems, and the necessity of including both human and non-human participants. On other occasions, much of the data and processes associated with what are called global problems are now held to operate below and before thresholds of human awareness. These kinds of formulations are increasingly hard to avoid. But the concept of the global and the related term globalisation are also contested, at least in some disciplines: it is argued that they cannot capture the intensities and unevenness of the variety of mobilities that crosscut the world (Sheller and Urry 2006). For this reason, some scholars prefer to explore the unevenness that might be introduced into an understanding of such mobilities by focusing on 'inter'-relations. This approach has more resonance with a consideration of the individuation of a problematic of the contemporary, and with an emphasis on the making or composing – rather than the finding – of problems.

Consider, for example, those scholars who have drawn on the idea of 'Asia as a method' (Chen 2010) to develop both an intellectual movement – Inter-Asia Cultural Studies (IACS) – and a methodology, Inter-Asia methodology. Inter-Asia as methodology involves visibilizing the normative frame that often has the 'West' as its key reference point, comparative work, and the identification of terms to become part of a broader conceptual framework. For Niranjana this includes what she calls the 'pressing' of concepts, that is a mutual or lateral interrogation of concepts with each other to reveal their interconnections and how they unevenly implicate simultaneously different Asian locations, the production of genealogies of the Asian 'present' (2013).

Infrastructural changes such as the increase in computational capacity, the growing availability of 'real-time' data and transformations in the ecosystem of data retrieval, have also contributed to a preoccupation with the relation of research to the future. In particular, these changes have stimulated discussion as to whether the future can somehow be brought into the present, that is whether the future can be not just predicted but in some sense anticipated. That is, for example, one way of understanding the changes in the calculation of risk associated with new techniques of statistical analysis (Amoore 2013). It is also linked to a resurgence of interest in design as a methodology and a concern with the performativity of method. The appropriation of design methodology is also associated with the making of artefacts of all kinds, including epistemic artefacts, synthetic materials or even smart cities. In research linked to smart materials, smart cars or smart cities, design methodology contributes to the creation of cognitive and epistemic artifacts that are held to have the capacity to modulate the present in what is described as real time, bypassing more human forms of governance. As epistemic infrastructures come to be equipped with real-time instrumentation, including actuators (Hayles 2017), and a growing number of storage and memory artefacts able to mobilise and articulate the potential-izing capacities of individuation, Simondon's claim that time itself is 'the expression of the dimensionality of the being as it is becoming individualised' (1992: 314) is being activated methodologically.

-ing

By focusing on form as in-formation, the proposal of the term compositional methodology is designed to draw attention to the uneven, non-linear temporalities of this methodological activation. The thinking behind the term was developed in the perhaps unlikely context of working on an *International Handbook of Interdisciplinary Methods* (2018; with co-editors Rachel Fensham, Alexandra Heller-McCrea, Angela Last, Mike Michael, and Emma Uprichard), the working shorthand for which was -ing! Contributors were asked to describe the *do-ing* of their chosen methods.

An inspiration for this approach was the artist Richard Serra's claim that 'Drawing is a verb'. His artwork *Verb List* (1967-68) serves as a kind of manifesto for this pronouncement. In pencil, on two sheets of paper, in four columns of scripts, the artist lists the infinitives of 84 verbs – *to roll, to crease, to fold, to store*, for example – and 24 possible states or conditions – *of gravity, of entropy, of photosynthesis, of nature* among others. In an interview, Serra says, '[t]he problem I was trying to resolve ... was: How do you apply an activity or a process to a material and arrive at a form that refers back to its own making?' (Garrels 2011). The art critic Rosalind Krauss suggests that the list describes Serra's own practice in terms of action that 'simply acts, and acts, and acts' (1985: 101). Serra himself draws attention to the relations in which the action that 'simply acts' takes place: he describes the list as a series of 'actions *to relate to oneself, material, place, and process*' (Buchloh 2000: 7; my emphasis).

The verbs in the list are transitive, that is, in linguistic terms, they can imply or express an object – to roll pastry, to crease paper, to fold metal, to store data, for example. This formulation – of implication or expression – is important; the verbs are not 'applied' to materials – they imply or implicate an object in a process, 'something/happening' as one of the contributors, Thomas Jellis, says in a discussion of experimenting (Jellis 2018: 53). This, perhaps, is why 24 examples 'of' a variety of objects, conditions or states are also included in the list by Serra: verbs are expressions *of* objects, conditions or states, and objects, states and conditions are the implication or expression *of* verbs.

Perhaps most significantly for a compositional methodology, *how* the verbs in Serra's list imply or express an object is a problem – indeed, it is 'the' problem. As he puts it, the problem is how to accomplish a form by or in doing. Drawing on this insight, and the contributions to the Handbook, the proposal I want to advance here is that the individuation of a problematic of the contemporary is an accomplishment of the do-ing of a method or methods, that is, it is an accomplishment of a practice in which there is a referral back to person, place, matter and process. For Serra, as for Simondon, that this accomplishment emerges as a form cannot be assumed: the operation of methods may or may not arrive at a form, have purchase on a question, or individuate a problem.

> '[n]ot having turns of phrase and modes of conjugation indicating processuality (like the English form -ing that indicates an action in the process of happening) available to him in his language, Simondon is to some extent constrained, in order to introduce dynamism into his thought, to invent a style.' (2013: 109).

In contrast to Serra's use of the infinitive form of verbs in *Verb List*, the approach introduced in the Handbook places emphasis on what are, in the English language, known as *gerunds*, that is, active present tense forms that may also function as nouns (a doubling that is made explicit in Serra's phrase, 'Drawing is a verb').[2] Put rather grandly, the Handbook's concern with methods as gerunds or -ings is intended to identify the potential of methods to

---

2 In English, this verb form typically ends in -ing, which is why our informal name for the Handbook was –ings! As Rachel Fensham, one of the co-editors of the Handbook, pointed out in discussion, other languages do not necessarily have the same verb forms. Indeed, as Combes points out in her influential explication of Simondon's conception of individuation, '[n]ot having turns of phrase and modes of conjugation indicating processuality (like the English form -ing that indicates an action in the process of happening) available to him

compose problems as interruptions of the (historical) present. That is, the aim of a compositional methodology is to emphasise the role of methods in the making of problems as an activation of the present: they are of interest insofar as they contribute to the determination of a situation as a problem, that is, 'a state of things in which something that will perhaps matter is unfolding amidst the usual activity of life' (Berlant 2008: 4). Just as Simondon says of a concept that it 'is neither *a priori* or *a posteriori* but *a praesentia*, because it is an informative and interactive communication between that which is larger than the individual and that which is smaller' (1992: 310), so we sought to describe interdisciplinary methods. Put rather more prosaically, the aim was to consider how methods might constitute some aspect of what is given, the present – in all its geo-political complexity – as a situation that has potentials that can be methodologically activated in specific, precise ways.[3]

How to realise this aspiration is described in a variety of ways in the Handbook. Matthew Reason says in his discussion of the method of drawing, '[d]rawing is at once immediate, and yet takes time':

> When I ask a participant to draw me a picture I am inviting a different dynamic than if I had simply asked them to talk. I do not expect them to respond instantly. Instead drawing imposes a slowing down, a pause for reflection in the returning to memories. (Reason 2018: 48)

Gail Davies and Helen Scalway say of the diagram,

> [...] it hasn't got a beginning, it hasn't got an end but nonetheless the incommensurable meanings are there, written in, but it hasn't got to have that linear structure of time. (Davies and Scalway 2018: 226)

---

in his language, Simondon is to some extent constrained, in order to introduce dynamism into his thought, to invent a style.' (2013: 109)

She continues: 'For all its subtlety, this style is nonetheless tangible, relying in large part on a specific usage of punctuation: it is thus not rare to see deployed, in a phrase composed of brief compositions connected with semicolons, all the phases of a movement of being or an emotion.' (ibid)

3 LaMarre, in his translation of Combes, uses the word 'givenness' in place of 'il y a' (which he also describes as a gloss on 'es gibt'), all phrases used to refer to 'being as such', in contradistinction to individuated being.

Alex Wilkie observes of speculating:

> Speculation, however, requires a shift in approach from analysing how probabilistic futures are manifested, managed and contested in the present – how actors imagine, model, predict, coordinate and in turn configure the future to the present – to the construction of adequate concepts and devices for exploring possible latent futures that matter. A word of caution is in order here, however: speculation is both prospective and retrospective. It applies as much to the politics of explaining past events (what might have been) as it does to the capturing of future possibilities (what might be). (Wilkie 2018: 347)

Catherine Ayres and David Bissel use the term suspending to describe the analytical potential of acknowledging the multiple durations present in an interview. They say,

> Different durations resonate at different times, sometimes immediately, and sometimes years after the initial encounter. Following Ingold's (1993) observations about the multiple co-existent temporalities of landscapes, we want to show how the interview "landscape" is steeped in the pasts and possible futures of researcher and researched alike, a site in which trajectories converge and transform. We want to revisit the interview event between Catherine and John to draw out "suspending" as a methodological intervention filled with theoretical, practical and ethical possibilities for thinking empirical encounters. (Ayres and Bissell 2018: 76)

Jussi Parikka says:

> As a method, digging opens up historically constructed material reality. It does not merely expose "ruins" but the multiple historical realities where material infrastructures have been layered, revealing different "distinctive temporalities and evolutionary paths" (Mattern 2015: 14). In this sense, digging opens the different temporalities that are all the time layered in infrastructures of cities, in media technological objects and in everyday situations. (Parikka 2018: 164-5)

Importantly, this activation of the present in the individuation of a problematic is not something that is a one-off, a discrete procedure, but rather, something that itself is typically conducted as part of a distributed and collaborative process in which a problematic individuates and is individuated. In other words, just as LaMarre notes in his discussion of technical individuation, 'because machines also exist in series and in ensembles, we also need to look at their phylogeny, at the relation between reproduction and transformation' (1999: 104), so also do we need to look at the distributed or differentiated reproduction of problems. To use a different, although related, set of terms, problems are always the methodologically induced property of distributed practical fields, themselves comprising researchers, methods, materials and media, connected to each other in time and space in diverse ways as part of the constantly changing epistemic infrastructure. Or, as Combes puts it, transductive unity is accomplished through 'a relative store of the "spacing out of being", its capacity for dephasing' (2013: 6).

To explore what all this might mean in relation to an individuation of a problematic of the contemporary let me return to the notion of process described by Simondon as transduction, and described by Whitehead as conformation–the interweaving of data, form, transition, and issue. Simondon says:

> By transduction, we mean a physical, biological, mental or social operation, through which an activity [of relation] propagates from point to point within a domain, grounding this propagation in the structuration of the domain, which is operated from place to place: each region of the constituted structure serves as a principle of constitution for the next region. (Simondon, quoted in Combes 2013: 6)

In this process, the operations of in-formation link the problematic and the associated milieu *and* ground the links between internal and external milieu. This is a difficult process to grasp but it is one way to acknowledge that a problem does not simply take place or individuate in an unchanging context, nor only to recognise that it is simply the context that is changing, but to acknowledge that the changing context is (also) changed by each action. One way of understanding this is through the notion of recursion, where recursion is understood as a repetition that sums as it sequences. In other words, the action that simply acts in the individuation of a problematic of the contemporary is not so simple at all, but, rather, compounds the problem. A paradigmatic example is free software, where the infrastructure (code) is self-grounded by the very collaborative effort that sets it in motion (Kelty 2008). In earlier work (Adkins/Lury 2011; Lury/Wakeford 2012) I used the term auto-spatialisation – adopted from the philosopher Gilles Châtelet (1999) – to describe this process of individuating a problem. From a compositional methodology point of view, the operation of individuation as auto-spatialisation can be taken to mean that a problem is always both a composite and compositional, that is, a form that is *of* the process of being *in*-formed. And this composition continually actualises potential as it makes relations between a problem and (the changing) context anew: a problem individuating is never fully solved but always in suspension.

Auto-spatialisation

Let me give some examples from a variety of disciplinary and interdisciplinary practices.

First, sampling. Zeilinger (2014) proposes that we recognise sampling as a methodological intervention, not simply as borrowing or stealing, but as a purposeful replacement of a recognizable original. He gives two cases. The first concerns a record of the Bobby Darin 1959 hit song *Dream Lover*, which as a result of the repeated playing of selected passages is scratched so that the needle gets stuck, repeating certain grooves: 'the jumping needle transforms the line "Dream lover, where are you – with a love oh so true?" into a loop that sounds like: "Dreamlo-lo-lo-lover where are yo-u-u-u-u [...]?"' (2014: 163-164). The effect of this stuttering, he says, subtly changes the line's connotation: 'The scratched record takes on the quality of a new utterance [...], and, from its inscribed involuntary repetitions and stammering, the listener may discern the longing for love, the insecurities, and the unfulfilled desires of a whole generation of listeners.' (2014: 164) The second case concerns the film *Alone*. Zeilinger says,

> [b]y my estimation, Arnold's *Alone* appropriates a total of around two minutes from several source films and, by inserting countless repetitions of sampled snippets, stretches the source material to roughly eight times its original duration. This intervention allows the filmmaker to focus on a number of arche-

typical character constellations (such as Father-Son, Father-Mother-Son, and Mother-Son-Love Interest) and to foreground in these constellations psychological issues that the cultural mainstream tends to gloss over. (2014: 164)

Through these two cases, Zeilinger argues that sampling can uncover, foreground, and repurpose the meanings of original materials. He shows how sampling can *return* the past to the present and the future. In making this argument he is very much concerned with the ways in which repetition is sequenced, that is, in the terms being developed here, how the organisation of auto-spatialisation can individuate a problematic in very particular ways:

When we sample, we do not necessarily produce anti-authoritative ruptures (that would be the legal action of the sampling artist as pirate); rather, sampling allows us to become part of circuits of meaningful repetition that can create new intimacies, new rapports between us, the original work, and the sampling piece itself. Sampling simultaneously dismantles and reinstates a work, an idea, or a unit. (2014: 169)

A second example comes from a study of Fluidity, the name of an open source computational code that is the key methodological resource of a large group of scientists at the Applied Modelling and Computation Group at Imperial College in London. In his ethnography of this group, Matthew Spencer describes the complex temporalities that are involved in the transformation in use of this code by scientists in different disciplines, working both independently and collaboratively, in syncopated rhythms with each other. He writes,

[r]esearch projects carry with them the whole weight of their past. While the trajectory of construction may move from a mathematical model of an analytical solution to a model of a well-studied experiment, the results of these previous stages become concretised in the apparatus as part of a testing system. When a scientist moves on to model something new, it is important to be assured that changes made in doing this have not undone earlier successes that built the foundation for the project. So as a test incorporated into the automated build and test suite, the earlier result will be run every time modifications are made to the code, ensuring that confidence from past success can still hold. [...] When a model is under active development, it is

never enough to cite validations and verifications that have been made in the past, because these have been made with respect to a different code. All past verification and validation accreted in the present system of research is thus carried forwards with current research projects, applied over and over again to every new iteration of the code. (Spencer 2013: 107)

This characteristic – the operationalisation of repetition in ways that actively engage and exploit a context that is itself changing – is an aspect of all individuation (including the individuation of a problematic), so Simondon would suggest. As Spencer acknowledges, scientific practice has always been repetitively distributed in space and time, but his study shows the newly enhanced methodological importance of the changing organisation of that repetition in a shared computational infrastructure *that is itself being deliberately changed* in a process of auto-spatialisation.

A third example relates to the use of computer-generated images (CGIs) in urban planning. Rose, Degen and Melhuish (2014) argue that rather than seeing them as still images, as static representations of urban space, they should be understood as interfaces circulating through a dynamic software-supported network space:

[...] the action done on and with CGIs as they are created takes place at a series of interfaces. These interfaces – between and among humans, software, and hardware – are where work is done both to create the CGI and to create the conditions for their circulation. (2014: 386)

Crucially, understood as interfaces, the circulation of a CGI is not secondary to its creation, but both a condition and a consequence of its methodological value in a process of auto-spatialisation.

A fourth example – perhaps the paradigmatic one – concerns the increasing importance attached to search in the conduct of re-search. David Stark observes a shift in the ways in which networks are transforming the processes of classification that are fundamental to many kinds of research (2011: 169). Things changed, he says, when the founders of Google reorganised search from a classificatory to a network logic:

[...] new social technologies exploit, radically in recombination, the three basic activities of life on the Web: *search, link, interact*. [... S]earch based on the structure of the links [...]. Interact based on the structure of searches [...] [L]ink based on the structure of the interactions. (2011: 171)

Stark emphasises the capacity of this new logic of search to correct a deficiency of methods that conceive databases as 'passive' and model search as information retrieval, that is, assume that the existing, often static, structure of an information resource contains all the relevant knowledge to be discovered. In contrast, Stark says, '[o]nce the vast databases are seen as an associative knowledge structure, the goal is to make them accessible as evolving knowledge repositories' (2011: 171). New categories emerge by treating users themselves as information resources with their own specific contexts. While the concept of category is not abandoned it is reconceived in relation to contexts produced in relation to circulation or movement:

> [...] short-term categories bring together a number of possibly highly unrelated contexts, which in turn create new associations in the individual information resources that would never occur with their own limited context. (Stark 2011: 173)

Each of these examples demonstrates the methodologically constitutive operation of auto-spatialisation, including, perhaps most significantly, practices of contexting that provide (unevenly) shared resource for the individuation of a problematic in terms of movement within and through changing milieu.

Rendition

To consider what is at stake in these practices for the individuation of a problematic of the contemporary let me introduce the artist Hito Steyerl's description of how the practice of film editing is currently being transformed. It is, she says,

> [...] being expanded by techniques of encryption – techniques of selection – and ways to keep material safe and to distribute information. Not only making it public, divulging or disclosing, but really finding new formats and circuits for it. I think this is an art that has not yet been defined as such, but it is,

well, aesthetic. It's a form. [...] Now it's not only about narration but also about navigation, translation, braving serious personal risk, and evading a whole bunch of military spooks. It's about handling transparency as well as opacity, in a new way, in a new, vastly extended kind of filmmaking that requires vastly extended skills. (Steyerl/Poitras 2015: 311)

Steyerl proposes that the question of how information is 'stored, secured, circulated, redacted, checked, and so on [... the] entire art of withholding and disseminating information and carefully determining the circumstances' is a 'formal decision'. She emphasises that this decision has an unstable temporality:

When I'm working with *After Effects*,[4] there is hardly any real-time play back. So much information is being processed, it might take two hours or longer before you see the result. So editing is replaced by rendering. Rendering, rendering, staring at the render bar. It feels like I'm being rendered all the time.

What do you do if you don't really see what you edit while you're doing it? You speculate. It's speculative editing. You try to guess what it's going to look like if you put key frames here and here and here. Then there are the many algorithms that do this kind of speculation for you. (Steyerl /Poitras 2015: 312)

In dialogue with Steyerl, the filmmaker Laura Poitras discusses the programme TREASUREMAP used by the US National Security Agency (NSA) to provide analysts with 'a near-real-time map of the internet and every device connected to it'. She suggests that at the core of the NSA's approach to data collection is a 'retrospective querying – how to see narrative after the fact' (Steyerl and Poitras 2015: 312).[5]

In the terms being developed here Steyerl provides a description of a specific form of auto-spatialisation – one which I suggest is increasingly dominant: rendition (see also Day and Lury 2017). Rendering or rendition is a term with many everyday as well as technical definitions, including: a per-

---

4 Adobe *After Effects* is a software tool for video compositing, motion graphics design and animation.

5 Following this line of thought, phenomena such as 'fake news' can be understood as epistemic artefacts of practices of rendition.

formance, a translation, an artistic depiction, a representation of a building executed in perspective, as well as meaning to return, to make a payment in money, kind or service, to pay in due (a tax or tribute) and, in legal terms, to transfer persons from one jurisdiction to another. The origin for all these uses of the term is the Latin *reddere*: 'to give back.' However, as Steyerl's description makes clear – and the current salience of the term 'forced rendition' also highlights – rendition can operate in ways that are deeply challenging to artistic practice. More widely, Steyerl's description of the changing conditions of editing provides a way to identify some of the issues facing anyone concerned with the individuation of a problematic of the contemporary. Among many others, these include: the 'auto' of auto-spatialisation; the composition of styles of reasoning; and transcontextualism and radicalizing contexts.

First, the 'auto' of auto-spatialisation. In relation to the changes in epistemic infrastructures described earlier, the provocation provided by my introduction of the notion of rendition makes visible the political importance of negotiating the tension between auto-as-autonomy and auto-as automatism. This concern is perhaps most evident in discussions of algorithms in general and machine learning in particular (Mackenzie 2017), since the evaluation of these methods relates to how they operationalise recursion (Fuller and Goffey 2012; Totaro and Ninno 2014). What is of concern is the kinds of control – the new kinds of normal, for example – that are established as recursion is used to 'organise heterogeneous material into a continuous, self-consistent pattern'. And while 'each recursive event is different, in terms of its scale, location in time, in the complications it may entail, and in terms of its place in relation to its nesting within other recursions or to those in which it is in turn nested', it is still by no means clear whether and how these methods are able to automate statistical induction to meaningful effect (Pasquinelli 2017).

This point leads onto the second issue – composite styles of reasoning. Simondon is keen to distinguish transduction from both deduction and induction. He says that, unlike deduction, transduction 'does not seek elsewhere a principle to resolve the problem at hand; rather, it derives the resolving structure from the tensions themselves within the domains' (1992: 315). And transduction is not comparable to induction, he says, 'because induction retains the character of the term of the reality as it is understood in the area under investigation – deriving the structures of analyses from these terms themselves [...] it only retains that which is positive, which is to say, that which is common to all the terms, eliminating whatever is singular' (1992: 315). Abduction – as it is commonly understood as the formation of explanatory hypotheses – has more affinity with transduction insofar as it is often explicitly defined in terms of inventiveness or creativity (Schurz 2008). However, the examples of auto-spatialisation outlined above suggest that transduction need not be restricted or reduced to any of deduction, induction or abduction, but that – in the individuation of specific problematics – all such styles of reasoning may play a role, in different ways, in different phases. Perhaps this is one way to think of a moving ratio?

Certainly the changes in epistemic infrastructures described above invite and support re-combinations of styles of reasoning: they allow, for example, for a variety of kinds of feedback, reciprocity and repetition. These, in turn, enable a more explicit engagement with what Simondon describes as the allagmatic dimensions of individuation, in which the concern is with description rather than explanation, or perhaps better, there are more ways to link description and explanation (Uprichard 2013; Mackenzie 2015). These include ongoing (epistemologically diverse and heterogeneously composite) experiments in anticipation (Ramírez and Selin 2014), speculation (as above and see Wilkie, Savransky and Rosengarten 2017), prototyping (Corsín Jiménez 2017), agent-based modelling, and simulation (Gilbert 2008; Gilbert and Troitzsch 2005). At the same time, the questionable epistemological status of, for example, 'narratives after the fact', the unstable evidential value of the possible rather than the probable, and concerns about whether and how anticipation comes to be associated with a pre-emption of the future all point to the political as well as epistemological dilemmas involved in the individuation of a problematic of the contemporary.

The third issue provoked by the notion of rendition and shared by the examples above can be understood in terms of the genus of syndromes that Gregory Bateson describes as transcontextualism. By this term, Bateson refers to a variety of cognitive tangles sharing common features, which he says are a result of the 'weaving of contexts and of messages which propose context – but which, like all messages, whatsoever, have 'meaning' only by virtue of context' (1972: 275-276). He says that most of these syndromes are not to be regarded as pathological (although his own thinking about the transcontextual relates to his attempts to understand schizophrenia), but should rather be understood as 'double takes' of a variety of kinds. Examples include when '[e]xogenous events may be framed in the context of dreams, and internal thought may be projected into the contexts of the external world' (1972: 200).

While Bateson's discussion of the implications of transcontextualism is largely confined to the level of the individual organism, his observations have considerable relevance in relation to the individuation of a problematic of the contemporary insofar as there is now an increased potential for contexts to be multiplied, and for heterogeneous cultures of contexting (Seaver 2015) to collide as well as for contexts to be equipped so as to be able to be interactive (Lury and Marres 2015).[6] As Bateson notes, there are a variety of ways to respond to or inhabit the transcontextual, including minimizing encounters with the transcontextual and actively resisting transcontextual pathways. However, neither way seems adequate at a time of 'radicalizing contexts'. As Antoinette Rouvroy and Thomas Berns point out, 'action based on the anticipation of individual behaviours could in the future be increasingly limited to an intervention on their environment, especially if the environment itself is reactive and intelligent, that is, if it collects data in real-time through multiple sensors, and shares and processes them to constantly adapt to specific needs and dangers, which is already the case at least during the significant part of life that individuals spend online' (2013: 172). Moreover, as both Bateson and Simondon acknowledge, (knowledge) propositions are always affective and have the potential to be pathological. Bateson writes,

> [p]sychologists commonly speak as if the abstractions of relationship (dependency, hostility, love, etc.) were real things which are to be described or expressed by messages. This is epistemology backwards: in truth, the messages constitute the relationship, and words like 'dependency' are verbally coded descriptions of patterns immanent in the combination of exchanged messages. (1972: 201)

---

6 For Cooley, communications provide for not only an extension, but also a possible multiplicity of environments. As a consequence, selection emerges as a formative, infrastructural dynamic in which 'a million environments solicit' the emergent individual (Cooley 1897: 23). A mundane example of this is the situation in which wifi networks solicit you (or your machine) to join them, the relative strength of their signals disrupting any continuity of context.

Conclusion

This discussion has addressed a subsidiary but integral dimension of Simondon's thinking: the individuation of a problematic. It introduced the term compositional methodology to draw attention to the methodological dimensions of this process, focusing on the temporalities involved in what was described as the activation of the present or the folding of process into practice as part of the individuation of a problematic of the contemporary. It further sought to demonstrate that the methodological aspects of the individuation of a problematic of the contemporary cannot be separated from the affective, moral and political aspects of this process by introducing the notion of rendition alongside the term auto-spatialisation.

In conclusion, I want to return to the understanding of the contemporary and its significance for an understanding of the individuation of a problematic. On the one hand, if we wish to avoid the present-ism of which Osborne speaks it is important to acknowledge Simondon's relation to the cybernetic theory that was emerging at the time of his writing (see Hörl 2012). He himself recognised the importance of doing so, distinguishing between what he considered to be his own qualitative interpretation of information and the quantitative understanding operationalised by Shannon and Weaver (Hayward and Geoghegan 2012). But still the question remains: Does Simondon's use of the concept of in-formation rely on an understanding that is insufficiently distinguished from that which is operationalised in the practices which are, very often, the object of study? That is, does his reliance on a notion of in-formation make him unable to challenge cybernetic thinking and practice and so make his understanding of individuation inadequate for understanding the contemporary? On the other hand, to avoid the pitfalls of a deterministic historicism, perhaps we need to recognise that the individuation of a problematic of the contemporary has potentials that can only be actualised in relation to the impersonal power of a shifting present (Esposito 2012). While being aware of the possibility of foreclosure of thought implied in rendition, the analysis above suggests that we should not retreat to the (un)certainties of a knowing subject, but rather assess what it means for uncertainty about to be distributed unevenly, and what conceptual personae – the idiot (Stengers 2005), the lurker (Goriunova 2017), the digital subject (Wark 2018), the machine learner (Mackenzie 2017) – might be adequate as we 'individuate, individuate ourselves and in ourselves'.

References

Adkins, Lisa/Lury, Celia (2011): 'Introduction: Special Measures.' In: Sociological Review 59/2, pp. 5-23. https://doi.org/10.1111/j.1467-954X.2012.02051.x

Amoore, Louise (2013): The Politics of Possibility: Risk and Security Beyond Probability, Durham, NC: Duke University Press. https://doi.org/10.1215/9780822377269

Ayres, Catherine/Bissel, David (2018): 'Suspending.' In: Celia Lury et al. (eds.), International Handbook of Interdisciplinary Methods, London: Routledge, pp. 76-80. https://doi.org/10.4324/9781315714523-9

Back, Les/Puwar, Nirmal (eds.) (2012): Live Methods, Oxford: Wiley-Blackwell.

Bateson, Gregory (1972): Steps to an Ecology of Mind, Chicago: University of Chicago Press.

Berlant, Lauren (2008): 'Thinking about feeling historical.' In: Emotion, Space and Society 1/1, pp. 4-9. https://doi.org/10.1016/j.emospa.2008.08.006

Blencowe, Claire (2015): 'The matter of spirituality and the commons.' In: S. Kirwan/L. Dawney/J. Brigstock (eds.), Space, Power and the Commons: The Struggle for Alternative Futures, London: Routledge, pp. 185-203.

Buchloh, Benjamin (2000): 'Process sculpture and film in the work of Richard Serra.' In: H. Foster/G. Hughes (eds.), Richard Serra, Cambridge, MA: The MIT Press, pp. 1-19.

Châtelet, Gilles (1999): Figuring Space: Philosophy, Mathematics and Physics, New York: Springer. https://doi.org/10.1007/978-94-017-1554-6

Chen, Kuan-Hsing. (2010): Asia as Method: Toward Deimperialization, Durham, NC: Duke University Press. https://doi.org/10.1215/9780822391692

Connelly, William E. (1999): Why I am not a Secularist, Minneapolis: University of Minnesota Press.

Cooley, Charles H. (1897): 'The process of social change.' In: J. D. Peters/P. Simonson (eds.), Mass Communication and American Social Thought Key Texts 1919-1968, Lanham, MD: Rowman & Littlefield.

Combes, Muriel (2012): Gilbert Simondon and the Philosophy of the Transindividual, Cambridge, MA: MIT Press.

Connor, Steven/Corby, Tom/Nafus, Dawn/Redler Haws, Hannah/Smith, Marquard/Teasley, Sarah, Numbers/Data: A Roundtable. In: Journal of Visual Culture 16/3, pp. 355-85. https://doi.org/10.1177/1470412917742083

Corsín Jiménez, Alberto (2014): 'The right to infrastructure: a prototype for urban source urbanism.' In: Environment and Planning D: Society and Space 32, pp. 342-62. https://doi.org/10.1068/d13077p

Corsín Jiménez, Alberto (ed.) (2017): Prototyping Cultures: Art, Science and Politics in Beta, Abingdon: Routledge. https://doi.org/10.4324/9781315529493

Davies, Gail/Scalway, Helen (2018): 'Diagramming.' In: Lury et al. (eds.), International Handbook of Interdisciplinary Methods, London: Routledge, pp. 209-227. https://doi.org/10.4324/9781315714523-31

Day, Sophie/Lury, Celia (2017): 'New Technologies of the Observer: #BringBack, Visualization and Disappearance.' In: Theory, Culture and Society 34/7-8, pp. 51-74. https://doi.org/10.1177/0263276417736586

Esposito, Elena (2013): 'Digital prophecies and web intelligence.' In: M. Hildebrandt/K. de Vries (eds.), Privacy, Due Process and the Computational Turn: The Philosophy of Law Meets the Philosophy of Technology, Oxford and New York: Routledge, pp. 121-42.

Esposito, Roberto (2012): Third Person: Politics of Life and Philosophy of the Impersonal, Cambridge, UK: Polity Press.

Flusser, Vilém (2014): Gestures, Minneapolis: University of Minnesota Press. https://doi.org/10.5749/minnesota/9780816691272.001.0001

Fuller, Matthew/Goffey, Andy (2012): 'Digital Infrastructures and the Machinery of Topological Abstraction.' In: Theory, Culture and Society 29/4-5, pp. 311-33. https://doi.org/10.1177/0263276412450466

Garrels, Gary (2011): 'An Interview with Richard Serra (2010).' In: Richard Serra Drawing: A Retrospective, Houston: The Menil Collection.

Gilbert, Nigel/Troitzsch, Klaus G. (2005): Simulation for the Social Scientist, Maidenhead: Open University Press.

Gilbert, Nigel (2008): Agent-Based Models, Thousand Oaks: Sage. https://doi.org/10.4135/9781412983259

Goriunova, Olga (2017): 'The lurker and the politics of knowledge in data culture.' In: International Journal of Communication 11, pp. 1-17.

Hayles, N. Katherine (2017): Unthought: The Power of the Cognitive Nonconscious, Chicago: University of Chicago Press. https://doi.org/10.7208/chicago/9780226447919.001.0001

Hayward, Mark/Geoghegan, Bernard D. (2012): 'Introduction and catching up with Simondon.' In: SubStance 41/3. https://doi.org/10.1353/sub.2012.0027

Hörl, Erich (2012): 'Luhmann, the non-trivial machine and the neocybernetic regime of truth.' In: Theory, Culture and Society 29/3, pp. 94-121. https://doi.org/10.1177/0263276412438592

Ingold, Timothy (1993): 'The temporality of the landscape'. In: World Anthropology 25/2, pp. 152-74. https://doi.org/10.1080/00438243.1993.9980235

Jellis, Thomas (2018): 'Experimenting'. In: C. Lury et al (eds.), Routledge Handbook of Interdisciplinary Methods, London and New York: Routledge, pp.53-7. https://doi.org/10.4324/9781315714523-4

Kelty, Chris (2008): Two Bits: The Cultural Significance of Free Software, Durham, NC: Duke University Press. https://doi.org/10.1215/9780822389002

Knorr Cetina, Karen (1999): Epistemic Cultures: How the Sciences Make Knowledge, Cambridge, MA: Harvard University Press.

Krauss, Rosalind (1985): The Originality of the Avant-Garde and Other Modernist Myths, Cambridge, MA: MIT Press.

LaMarre, Thomas (2012): 'Afterword: Humans and Machines'. In: Muriel Combes (ed.), Gilbert Simondon and the Philosophy of the Transindividual, Cambridge, MA: MIT Press. pp. 79-108.

Lury, Celia/Wakeford, Nina (eds.) (2012): Inventive Methods: The Happening of the Social, London: Routledge. https://doi.org/10.4324/9780203854921

Lury, Celia/Marres, Noortje (2015): 'Notes on objectual valuation.' In: M. Kornberger/L. Justessen/A. Koed Madsen/J. Mouritsen (eds.), Making Things Valuable, Oxford: Oxford University Press. https://doi.org/10.1093/acprof:oso/9780198712282.003.0012

Lury, Celia/Fensham, Rachel/Heller-Nicholas, Alexandra/Lammes, Sybille/Last, Angela/ Michael, Mike and Uprichard, Emma (eds) (2018): Routledge Handbook of Interdisciplinary Methods, London and New York: Routledge. https://doi.org/10.4324/9781315714523

Lury, Celia (forthcoming): Problem Spaces: Why Methodology Matters, Cambridge, UK: Polity.

Mackenzie, Adrian (2002): Transductions: Bodies and Machines at Speed, London: Bloomsbury Academic.

Mackenzie, Adrian (2015): 'The production of prediction: What does machine learning want?' In: European Journal of Cultural Studies 18/4-5, pp. 429-45. https://doi.org/10.1177/1367549415577384

Mackenzie, Adrian (2017): Machine Learners: Archaeology of a Data Practice, Cambridge, MA: MIT Press. https://doi.org/10.7551/mitpress/10302.001.0001

Marres, Noortje/Guggenheim, Michael/Wilkie, Alex (2018): Inventing the Social, Manchester, UK: Mattering Press.

Mattern, Shannon (2013): 'Infrastructural Tourism', online: https://placesjournal.org/article/infrastructural-tourism/ https://doi.org/10.22269/130701

Murphy, Michelle (2017): The Economization of Life, Durham, NC and London: Duke University Press.

Niranjana, Tejaswini (2013): Introduction to Genealogies of the Asian Present: Situating Inter-Asia Cultural Studies, online: https://www.academia.edu/14934670/Introduction_to_Genealogies_of_the_Asian_Present_Situating_Inter-Asia_Cultural_Studies.

Osborne, Peter (1995): The Politics of Time: Modernity and Avant-Garde, London/New York: Verso.

Parikka, Jussi (2018): 'Digging.' In: Celia Lury et al. (eds.), International Handbook of Interdisciplinary Methods, London: Routledge, pp. 164-68. https://doi.org/10.4324/9781315714523-25

Pasquinelli, Matteo (2017): 'Machines that morph logic: Neural networks and the distorted automation of intelligence as statistical inference.' In: Glass Bead Journal, Site 1, Logic Gate: The Politics of the Artifactual Mind.

Rabinow, Paul (2009): Marking Time: On the Anthropology of the Contemporary, Princeton: Princeton University Press. https://doi.org/10.1515/9781400827992

Ramírez, Rafael/Selin, Cynthia (2014): 'Plausability and probability in scenario planning.' In: Foresight 16/1, pp. 54-74. https://doi.org/10.1108/FS-08-2012-0061

Reason, Matthew (2018): 'Drawing.' In: Celia Lury et al. (eds.), International Handbook of Interdisciplinary Methods, London: Routledge, pp. 47-52. https://doi.org/10.4324/9781315714523-3

Rheinberger, Hans-Jörg (2010): An Epistemology of the Concrete: Twentieth Century Histories of Life, Durham, NC: Duke University Press. https://doi.org/10.1215/9780822391333

Rose, Gillian/Degen, Monica/Melhuish, Clare (2014): 'Networks, interfaces, and computer-generated images: learning from digital visualisations of urban redevelopment projects.' In: Environment and Planning D: Society and Space 32, pp. 386-403. https://doi.org/10.1068/d13113p

Rouvroy, Antoinette/Berns. Thomas (2013): 'Algorithmic governmentality and prospects of emancipation.' In: Réseaux 177, pp. 163-96. https://doi.org/10.3917/res.177.0163

Schurz, Gerhard (2008): 'Patterns of abduction.' In: Synthese 164, pp. 201-34. https://doi.org/10.1007/s11229-007-9223-4

Seaver, Nick (2015): 'The nice thing about context is that everyone has it.' In: Media, Culture and Society 37/7, pp. 1101-9. https://doi.org/10.1177/0163443715594102

Serra, Richard (1967-68): Verb List. https://www.moma.org/collection/works/152793

Sheller, Mimi/Urry, John (2006): 'The new mobilities paradigm.' In: Environment and Planning A: Economy and Space 38, pp. 207-26. https://doi.org/10.1068/a37268

Simondon, Gilbert (1992): 'The genesis of the individual,' In: J. Crary/S. Kwinter (eds.), Incorporations, Cambridge, MA: MIT Press, pp. 297-319.

Simondon, Gilbert (2017): On the Mode of Existence of Technical Objects, Minneapolis: University of Minnesota Press.

Simone, Abdoumalique (1994): In Whose Image? Political Islam and Urban Practices in Sudan, Chicago: University of Chicago Press.

Spencer, Matthew (2013): Reason and Representation in Scientific Simulation, PhD thesis, Goldsmiths College.

Stark, David (2011): The Sense of Dissonance: Accounts of Worth in Economic Life, Princeton/London: Princeton University Press.

Stengers, Isabelle (2005): 'The cosmopolitical proposal.' In: Bruno Latour/Peter Weibel (eds.), Making Things Public, Cambridge, MA: MIT Press, pp. 994-1003.

Steyerl, Hito/Poitras, Laura (2015): 'Techniques of the Observer: Hito Steyerl and Laura Poitras in conversation.' In: Artforum, May, pp. 306-17.

Totaro, Paulo/Ninno, Domenico (2014): 'The Concept of Algorithm as an Interpretative Key of Modern Rationality.' In: Theory, Culture and Society 31/4, pp. 29-49. https://doi.org/10.1177/0263276413510051

Uprichard, Emma (2013): 'Describing description (and keeping causality): The case of academic articles on food and eating.' In: Sociology 47/2, pp. 368-82. https://doi.org/10.1177/0038038512441279

Wark, Scott (2018): 'The Subject of Circulation: On the Digital Subject's Technical Individuations.' In: Subjectivity 12, pp. 65-81. https://doi.org/10.1057/s41286-018-00062-5

Whitehead, Alfred North (1968): Modes of Thought, New York: The Free Press.

Wilkie, Alex (2018): 'Speculating,' In: Celia Lury et al. (eds.), International Handbook of Interdisciplinary Methods, London: Routledge, pp. 347-51. https://doi.org/10.4324/9781315714523-49

Wilkie, Alex/Savransky, Martin/Rosengarten, Marsha (2017): Speculative Research: The Lure of Possible Futures, Abingdon/New York: Routledge. https://doi.org/10.4324/9781315541860

Zeilinger, M. J. (2014): 'Sampling as Analysis, Sampling as Symptom: Found Footage and Repetition in Martin Arnold's Alone. Life Wastes Andy Hardy.' In: D. Laderman/L. Westrup (eds.), Sampling Media, Oxford: Oxford Scholarship Online, pp. 159-71. https://doi.org/10.1093/acprof:oso/9780199949311.003.0012

From Critique to Problems and the Politics of the In-act with Bergson, Deleuze and James

Christoph Brunner

> Taking down our critique, our own positions, our fortifications, is self-defense alloyed with self-preservation. That takedown comes in movement, as a shawl, the armor of flight.
>
> We run looking for a weapon and keep running looking to drop it. And we can drop it, because however armed, however hard, the enemy we face is also illusory.
> (Harney/Moten 2013: 19)

Introduction – the problem of critique

I want to start with a productive paradox (or, a problem): In their work on *The Undercommons* Stefano Harney and Fred Moten begin their exploration of collective forms of resistance from a postcolonial, post-structuralist and post-operaist perspective with the problem of critique. They tie such a conception of critique to politics as it emerges with the process of building enclosures (as governable or knowable entities) in the process of settler colonialism: 'Politics is an ongoing attack on the common' (2013: 17). Such politics based on enclosure mobilize critique as an instrument that is representative of institutional power, as a form of positioning in defense of the enclosure (read as domain, discipline, institution). At the same time, 'critique lets us know that politics is radioactive, but politics is the radiation of critique [...]

Critique endangers the sociality it is supposed to defend' (2013: 19). By staging this double-edged impasse of critique as necessary and radioactive, the authors erupt the idea of a politics of enclosure and its defense of critique through the image of allied movements: the taking down of critique as a 'movement' as the 'armor of flight', 'looking for a weapon and keep running looking to drop it' (2013: 19). These minor gestures (Manning 2016) of movement, flight, the looking for and dropping of a weapon for self-defense all contest a critical thinking based on positions and a politics of enclosures.

The movement of flight, a concept well-known from the works of French feminist philosopher and poet Hélène Cixous (1976) and those of Gilles Deleuze and Félix Guattari (1987), underscores a general activity, a resonance of tendencies, a movement that relates to other movements. Flight and escape as derived from the French notion *fuite* in the works of Deleuze and Guattari take on multiple meanings: 'Both words translate *fuite*, which has a different range of meanings than either of the English terms. Fuite covers not only the act of fleeing or eluding but also flowing, leaking, and disappearing into the distance (the vanishing point in a painting is a *point de fuite*)' (Massumi in Deleuze/Guattari 1987: xvi). A takedown of critique as a movement of flowing, leaking and disappearing from politics and its capturing critique, radically alters the conception of political practice understood through a logic of oppositions. Such oppositional politics are based on practices of identification and order, they are reflexive and built on a casting of the real and truth built on common sense. The practice of forming enclosures, of naming and identifying fixes positions and overcodes the actual movement. It separates past from present and future, putting them into causal relations. The takedown of critique, as movement, disrupts the natural sense of how to seize a situation in all its complexity by putting things in place and establishing orders, enclosures and domains. Understanding the constitution of the real as based on non-linear movement, that is, activity, allows the real to be understood as the realm where actual problems occur.

The focus on movement, on flight as foundational activity of existence, stages politics as a question of continuous difference and critique as a way of tapping into the process of continual differentiation that resists terminal enclosure. This article will explore how to resist a politics that instrumentalises critique built on enclosure, common sense and presumed opposition, proposing an affirmative engagement with the invention of problems instead. According to French philosopher Henri Bergson the refutation of a politics of critique built on enclosure and common sense can only be contested by the constitution of *real* problems. Such problems, Bergson suggests, engage in a field of movements rather than entities resonating with each other to form what constitutes the real. While Moten and Harney are explicitly drawing on a variety of theories and references, including voices critical of the Western philosophical tradition, their emphasis on movement and flight links to a reconceptualisation of time as a colonising and colonised concept in modernity.[1] One might think of Afrofuturism's notion of the future, which is anything but a transcendent imaginary and rather a multiplicity of 'counter-memories' of the future stalled in the present (Eshun 2003: 288). Such futures are the movements of the takedown of critique Harney and Moten point at. The dropping of the weapon means not to succumb to the temporality of the present but to engage in a 'resistance to the present' (Deleuze/Guattari 1994: 108; Stengers 2010). The present to be resisted is one of a reactive mode of critique, of a temporality in which critique knows its outcome in advance of its utterance. It is Bergson who was most explicit in his assertion that most of modern Western philosophy has misunderstood truth as what builds on common knowledge and its orders, rather than something that needs to be sought after in experience.

The problem of critique is its mooring in a past that it claims to know and from which it stages its attack in order to colonise the future in a self-righteous manner. Harney and Moten do not refuse critique in general but ask how critical practice can take shape while not knowing in advance what the enemy might look like, or from which position of critique one speaks. The question of a politics built on movement not on enclosure immediately becomes a time-sensitive concern beyond linear succession, asking what constitutes a political act beyond a negative mode of critique. An affirmative practice of problematisation, 'dramatizing the creation of problems', becomes a liberating act from the fetters of a dominating and dominated present (Stengers 2019: 1).

In a first section the text will situate Henri Bergson's notion of critique and clarify his own thinking as a pragmatist. Exploring, secondly, the different yet mutually resonating ways of problematising the concept of common sense in Bergson, Gilles Deleuze and William James, I will expose their rad-

---

[1] On the relation between coloniality and modernity, see for instance Aníbal Quijano (2007). In relation to time and colonisation, see Mark Rifkin (2017).

ical intervention into the philosophy of knowledge and experience (During 2004) leading towards a movement thinking able to state real problems. The creation of problems, however, is neither an affair based on a formerly agreed conception of common sense nor of language but of *intuition*. The third part will treat Bergson's particular method of intuition as an alternative to critical philosophy based on common sense. Intuition as the method of stating real problems, takes movement or duration as the grounding operation which enables a conception of the real as compositions of time and space. The engagement with such a real from the perspective of human activity becomes an act of invention. Similar to afrofuturist temporality, it generates novel and singular perspectives capable of resisting the present based on common sense. The final section engages with the compositional ground of intuituton's activity of stating problems which are tendencies as the minimal existence of the real. Making tendencies the only thing that can be known turns the emergence of the real into a polyrhythmic dynamism. From here the notion of the in-act will allow to distinguish real encounters with tendencies' movement from acts based on the prefixed couplings of linear causation such as before–after, subject–object or present–absent. Politics, as I want to suggest, relates to what one considers an act capable of 'producing [...] lines of singularity, its own cartography, in fact, its own existence' (Guattari 1996: 136). Such an act is not of a single being but traverses vast distances and thus draws novel engagements with the real as the realm of activity.

False and real problems – from metaphysics to pragmatism

The question of the problem, and related to that of false problems, defines Bergson's critique of the negative as a category of division. Bergson's reservations about the classic metaphysical stance are clearly stated by Gilles Deleuze: 'His fundamental criticism of metaphysics is that it sees differences in degree between a spatialized time and an eternity which it assumes to be primary [...]: All beings are defined on a scale of intensity, between the two extremes of perfection and nothingness' (Deleuze 1988a: 23). At the same time Bergson himself clarifies that he has no intention of giving up metaphysics, but would rather develop 'a truly intuitive metaphysics, which would follow the undulations of the real' (Bergson 1946: 29). The problem of metaphysics as philosophical practice 'leaves no room to force metaphysics to speak of extrabeing' in the overall conception of the real as Foucault states (Foucault 1998: 347). *Extra-Being* is outside of the divide between being and Being, the false problem confusing a substantialist account of matter and its successive persistence over time. Bergson, as taken by Deleuze, would push metaphysics to speak of such extra-Being. Extra-Being describes a 'minimum of Being common to the real, the possible, and the impossible' and thus a domain outside of space and time, while informing both (Deleuze 1990: 180). Rather than wondering whether Bergson's philosophy would align with empiricism or positivism, the notion of Extra-Being as introduced by Deleuze, positions him as the proper philosopher of the virtual as existence outside of Being and being. The virtual as a domain of existence in tendencies guides the production of the real while neither reducing it to a given (data) nor to being able to fully abstract it in consciousness. Extra-Being is the realm where tendencies relate and thus shape the ground of experience from which perceptual events emerge. It is in this sense that Deleuze underlines that problems need to be considered 'as ideal "objecticities" possessing their own sufficiency and implying acts of constitution' instead of being inferred from anything prior or deduced through logic and reason (1994: 159). It is an 'objecticity' of the event as its very own mode of becoming expressive in actualisation. The temporality of extra-Being is ideal, because it does not need to actualise in order to be real – it has purchase in the real. A proper problem consists of embracing a real beyond actuality. However, extra-Being presumes no beyond the real but postulates an immanent temporality that is utterly untimely to both the past and the present, rendering them both as invested in future potentialities. In the entry quote the takedown of critique in flight is paralleled with the need for self-preservation, the looking for a weapon paired with the dropping of it. These acts are not opposites or contradictory, they rather constitute a politics of the real as tensed field of relating tendencies.

For Bergson, false problems relate to a mistaking of differences in kind for differences in degree. Questions such as 'Why is there something rather than nothing, order rather than disorder?' (Deleuze 2004: 25) are constitutive of false problems. Why? Because such questions pose a problem in the image of the negative, whose refusal Deleuze attributes to Bergson's 'repudiating critical philosophies' (2004: 23). The image of the negative, the lack, or the opposed would only ever contend itself with systems of order based on differences in degree (Deleuze 1988a: 17–20). Such differences are mere placements of duration in space, of a substantialism that knows where to put things and how to tell this from that. Real problems, on the contrary, ask 'Why this rather than something else? Why this tension of duration? Why this speed rather than another?' (1988a: 25). For Bergson, the emphasis on the problem resides in the paradox beyond the binary as the negative and thus a dominant image of thought. Such an image, as Deleuze develops throughout his entire work, lacks a proper account of the real as productive of 'encounters forcing us to think' (1994: 139). Similarly, Bergson writes 'philosophy, thus understood [...] will have no difficulty in explaining everything deductively, since it will have been given beforehand, in a principle which is the concept of concepts, all the real and all the possible' (1946: 34).

The figure of the negative is potentially the most common conception of critique that is at stake for Bergson and Deleuze – and in their aftermath Moten and Harney. The false problem, as the one that always constitutes an identity in the image of another, places these oppositions into a perpetual loop of classifications and orders ignorant of Extra-Being as the actual ground of emergence. These orders are helpful as orientations – they confirm and comfort but they do not leap into unknown territories. It is here, in the naming of false problems, that one of Bergson's most rebellious traits comes to the fore. If 'truth and creation are reconciled at the level of problems', problems replace the traditional logic of concepts and theories as *prior to* or deduced *from* experience (Deleuze 1988a: 15; During 2004: 19). In a very different register, which is more Deleuze's than Bergson's, false problems disregard the real according to singularities, turning each instant into a moment of particulars rather than accounting for their differential nature. False problems are problems operating by degree or intensity, while real problems only ever operate by the differentials expressed through singularities or singular points. This means that there is a uniqueness in each expression or manifestation but not only to the matter formed, but also to that formed in relation to its past and its future. Accounting for the process of becoming rather than placing beings into space means to radically rethink what critique and analysis mean for philosophical practice – but also for political practices. It requires to take Extra-Being into account as an affirmation of a time beyond order and enclosure into a reductive present.

Empiricism aligns with a projection of time into space, a positioning of sorts. Idealism points at the primacy of duration over space, thus foregrounding movement rather than substance or position. Going back to the initial quote of Moten and Harney, one finds such a productive paradox of processual thinking beyond a terminal conception of critique. To take down one's critique as self-defence and self-preservation, looking for a weapon while looking to drop it, the illusion of the enemy, these are tension and variations of speed rather than finite acts. In fleeing they manifest their existence as extra-Being, not a mere surplus or excess, but a different register of opening encounters that force us to think. The question of the problem is then how to engage with or stage such encounters, and how to account for their singular and enduring, yet continuously differentiating, qualities.

While more recent philosophical debates have delivered insights into Bergson's concept of the problem in relation to the history of philosophy, discussing its difference from epistemology and positivism in the French tradition of the 19th and 20th centuries (During 2004, Bowden 2018), Bergson himself clarifies his own position in his praise for early pragmatist philosopher William James.[2] In a preface to a translation of James' work on pragmatism Bergson states that real problems emerge when 'we confine ourselves purely and simply to what is given us by experience' (Bergson 1946: 249). However, this is not a positivist or empiricist stance, but a radical empiricism in the Jamesian vocabulary (James 1996) or a transcendental empiricism

---

[2] For the sake of clarification: Both articles by During (2004) and Bowden (2018) are crucial for the thinking of Bergson's notion of the problem and the writing of this chapter. However, their adherence to the French philosophical tradition requires further critical inquiry if one takes the more recent developments between 20th century French philosophy and early North American pragmatism into consideration (see for instance Savransky in this volume). Another important analysis, even though too close to the gestures of common sense as refused by Deleuze, relates Deleuze as a reader of James and relayed through Bergson (Madelrieux 2015: 89-91). The reading of Madelrieux exposes the philosophical gesture which Bergson and Deleuze would refute as false problems, when claiming that 'in three different and complementary ways, Deleuze misunderstood pragmatism. He misunderstood it firstly in that he assimilated pragmatism to pluralism. He missed it a second time since he borrowed the definition of pluralism from Bertrand Russell and not from William James. And he missed it a third time because his own version of pluralism does not stand up to the pragmatist method for making ideas clear' (2015: 89). From the get-go the article presents itself a severe misunderstanding of the Deleuzian philosophical project which affirms rather than criticizes ideas beyond a presumed common sense established by traditional philosophical reason. For a quite diverse exploration of the resonances between Deleuze and different strands of early and later pragmatist strands, see Bowden, Bignall, Patton 2015.

in Deleuze's work (1994). Bergson describes this radical empiricism attentive to the infinite nature of existence:

> While our intelligence with its habits of economy imagines effects as strictly proportioned to their causes, nature, in its extravagance, puts into the cause much more than is required to produce the effect. While our motto is *Exactly what is necessary*, nature's motto is *More than is necessary* – too much of this, too much of that, too much of everything. Reality, as James sees it, is redundant and superabundant [...] there are no sharply drawn situations; nothing happens as simply or as completely or as nicely as we should like; [...] things neither begin nor end; there is no perfectly satisfying ending, nor absolutely decisive gesture, none of those telling words which gives us a pause: all the effects are spoiled. (Bergson 1946: 249, emphasis in the original)

Following the abundant character of experience, Bergson further outlines what will lead towards his very own conception of problems and the method of intuition as a technique of problematisation. He refers to James' well-known attestation that relations need to be experienced as real as the things related, and adds that such relations are directly observable as 'the things and facts themselves' (1946: 250).[3] The acknowledgement of relations' facticity resonates strongly with Bergson's own claim that 'one must get back into duration and recapture reality in the very mobility which is its essence' (1946: 35). Relation is not an entity but a movement or trajectory, a tendency, of which many in attunement form reality.

The consequence of such a view, as outlined by Bergson, requires *a complete reversal of the image of thought that philosophy held of reality up until then*. Conceiving of reality as 'no longer finite or infinite, but simply as indefinite' renders 'reason [...] less at ease in a world where it no longer finds, as in a mirror, its own image. And certainly the importance of human reason is diminished. But the importance of man himself – the whole of man, will and sensibility quite as much as intelligence – will thereby be immeasurably en-

---

3 From James' *Essays in Radical Empiricism* (1996: 42): 'To be radical, an empiricism must neither admit into its constructions any element that is not directly experienced, nor exclude from them any element that is directly experienced. For such a philosophy, *the relations that connect experiences must themselves be experienced relations, and any kind of relation experienced must be accounted as "real" as anything else in the system.*'

hanced!' (1946: 250-251). Such an immersion into the relational fabrication of the real, qua *Extra-Being*, cuts across prior orders of enclosure, the extrapolation of human consciousness and embodiment *in* an empirical world, and puts duration at the centre of its conception. Why? Because it is through duration that *things* become different, not only from each other in a world of material experience, but also in relation to themselves, their own genesis. Bergson's embracing of James' conception of experience is not a mere acknowledgement of a world of experience much vaster than the human scope of sufficient reason might want to admit and capture, but also a plea for a more precise account of such a reality in its complexity and texture which common sense, as I will explore now, always accounts for insufficiently.

Beyond common sense

Bergson's critique of critical philosophy operates by affirmation and not negation, and thus requires a different mode of thinking about problems in their capacity to refuse a commonsensical agreement over norms and judgement. Unpacking the meaning of common sense leads us towards the intuitive method as the key junction between an actively self-affecting world and the composition of a thinking and acting subject. One could say, rather than deciding between empiricism and idealism, Bergson's speculative metaphysics is deeply rooted in a specific pragmatist understanding of experience and a refutation of a theoretical common sense.

A critique of common sense appears to define not only a particular shaping of the notion of the real based on duration and relations as facts but also functions as a major point of conceptual confluence between Bergson, James and Deleuze. For Deleuze, common sense is 'a moral or orthodox' image of thought tied to good sense (1994: 132). Common sense cannot conceive of paradoxes as problematising – the paradox here being the infinite character of experience, which is not a mere excess but a doubling of the very processes of encounters with *a problem*. In that sense, a problem does not appear out of thin air – it is fabricated, a constitutional act that takes hold of a singular situation. In the process of fabrication, a problem puts existence on the line, or to the test and renders it into a tensed field of resonant yet heterogeneous tendencies (Stengers 2019). It makes the situation of problematic emergence 'pointy' (Massumi 2015: 126), actualising its singular characteristics by shifting its emphasis to its very limit. This limit-character that problematisation foregrounds is the very act of becoming itself, at the limit, or inhabiting the limit.[4] Conceiving of the problem as a paradox means to emphasize its singular logic of infecting the real through its movement, its way of continuously referring to its specific mode of problematizing.

The rule of the paradox is what Deleuze poses against the allied repercussions of good sense and common sense in *The Logic of Sense*. The paradox is a reversal of common sense and good sense, it turns them upside down, queers them out of their operational alliance and plants the problematic amidst their impoverished accounts of the real. Good sense, the way Deleuze casts it, is unidirectional orientation from the 'most differentiated to the least differentiated' (1990: 75). In doing so, it generates an order of time, where the most differentiated is the past and the least differentiated is the future, thus colonising the present as oriented in that arrow of time. Such a concept of unique direction constitutes an image of thought whose orientation is *foresight* (1990: 75–76). One can glimpse how the directed orientation of the present under the 'principle of a unique sense' aligns with Bergson's critique of the subsumption of duration under space, turning duration into a reduced conception of a forward-moving present (1990: 76). The first major critique of this unique direction of time occurs through the assertion that the paradox consists of another *sense* moving into both directions simultaneously, thus extending into the past and the future while infinitely subdividing the present (see Deleuze 1989: 81). An example of the paradox operation of duration would be the use of polyrhythmic patterns in electronic dance music. While there is a driving beat (mostly 4/4) for which such music is best known, the temporal complexity of more sophisticated productions occurs through the layering of soundscapes, up to the level of granular sonic fragments. The ground beat could be seen as an utterly chronological driver of such music. On the contrary, I would conceive of it as the a-temporal ground through which different sonic elements as temporal patterns can move into specific constellations which are heard and felt. Based on the infinite a-temporality of the beat different sound events revolve and merge across this sur-

---

4 Deleuze points to such a concept of the limit in the 'Fifteenth series on singularities' in *The Logic of Sense*, referring to Simondon's own conception of singularity and his understanding of individuation as a process of ontogenesis. The temporal or duration nature of the limit defines a key link to the notion of the problem in Bergson (Deleuze 1990: 100–108).

face of the beat, and thus co-composing the musical experience. The paradox here is in the event of music in resonance but beyond synthesis, contracting temporalities on an infinite plane of the beat as its surface. The problem of multilayered sonic experience occurs through the durations that are constantly contracted without unifying, thus making space an element of time and not the other way around.

In Bergson's writings, the doubling of time refers to a spatialised time, which is matter, and a differential time of duration, found in memory. The paradox is, while duration has been falsely subsumed as spatialised in the conception of the real or the present, matter itself lends itself to such a temporality while actually being derived from duration. It is not the case that one has to solely align with duration to distill the true nature of differences in kind. In experience, we are confronted with mixtures and composites. The fabrication of a problem occurs when matter is directly related to memory, that is, its differentiation relays through the past (in general) that is simultaneous with the present. This simultaneity of present and past is crucial in order to account for the mixed states of existence that are neither just spatial nor temporal but contain two multiplicities – of differences in degree and of kind (see Bergson 1910: 110).

The doubling of time in Deleuze goes hand in hand with the doubling outlined in *Matter and Memory* (Bergson 1988). Therein Bergson clarifies the two fundamental differences at the heart of each existing as a mixed composite of both a spatialised time expressed in matter and a differential time inhering or insisting through duration or memory. The struggle over the present, the things and states of affairs, is a false struggle for Bergson and Deleuze as long as matter, spatialised time and differences in degree or intensity dominate the concept of experience. Put differently, as long as things are conceived in their substantialist casting as mere givens, they overcode the genetic nature that defines their differential essence. What counts, according to Deleuze and Bergson, is how 'things' differ in relation to themselves – this is the untimely temporality of becoming or extra-Being, which cannot be subsumed under the time of spatialised matter in Chronos.

The resonances with James cannot be underestimated here. When pointing at experience's abundant nature, Bergson agrees with James, manifesting a critique of philosophical reason of common and good sense content with the present's reductive representation and spatialisation rather than embracing a mode of encounter inclusive of the multiple durations exceeding such a present. The *real* becomes the terrain for the invention of real problems when they are posed in a way that they attend to the elements that sidestep the present without being absent from it.

Good sense, to maintain such an image of thought tied to foresight, requires another operation, which Deleuze attributes to common sense. Common sense is a 'faculty of identification that brings diversity in general to bear upon the form of the Same' (1990: 78). It defines the capture and enclosure of the predictive politics of critique that Harney and Moten problematise. In that way, common sense and good sense are the operations of an image of thought that is the constitution of a real solely based on a human-centered experience and consciousness. In its operation of foresight, good sense installs a temporal regime which allows it to colonize the future from the vantage point of the present. Common sense, on the other hand produces systems of resemblance and derivation without accounting for the real differences in kind based on duration.

Both Bergson and James link their refusal of common sense to a notion of economy. Bergson writes about the 'habits of economy', meaning the rationalised logic operating in critical philosophy ready to judge and classify according to order (1946: 249). James, for his part, uses the term as the 'triumph of economical thought' expressed in laws derived from scientific measurement. Such an economy, however, is not enough to account for reality. Accordingly, James claims: 'Profusion, not economy, may after all be reality's key-note' (1963: 85). Towards the end of his lecture on *Pragmatism and common sense*, he reiterates his suspicion of such an economic reason: 'Its [common sense's] categories may after all be only a collection of extraordinarily successful hypotheses [...] by which our forefathers have from time immemorial unified and straightened the discontinuity of their immediate experiences, and put themselves into an equilibrium with the surface of nature satisfactory for ordinary practical purposes that it certainly would have lasted forever.' (James 1963: 85)

James' insistence on the discontinuity of immediate experiences signals a crucial political quest outlined in relation to philosophy as a practice that potentially engages with or encounters problems. Approaching things with respect to their differentiating qualities, that is, their duration, means to account for the discontinuities of immediate experience as singularities beyond their discrete measures. It is the 'fissures and cracks' that co-ordinate a relational becoming, not an essence (Deleuze/Guattari 1987: 224). And it is the critique of such an economy that foregrounds the problem as a key concept – to resist a certain present and to 'experimentally think with the "situational provocation" of the present' (Stengers 2019: 2).

If good sense, as Deleuze writes, 'determines the contribution of the faculties in each case, while common sense contributes the form of the same' then the political question of the paradox and the problem is, how to reenter experience's engagement with Extra-Being. As I have tried to show, good sense and common sense are primarily temporal operations. They align disparate and divergent temporalities into coherent order of moments which lead from a past towards the future while inhabiting the present. Underneath this reduction, the actual compositional activity of duration underlines that everything already moves and that it is movement which renders emergence possible. What occurs through a problem is difference, a 'difference which forces us to think' (Deleuze 1994: 136). Problems engage the real beyond 'recognition, today or tomorrow' and tie the process of an embodied experience into the overall welter of experience expressive of activity (ibid).

The pragmatist method of posing or stating problems then requires a certain adjustment to the situations to which these problems provide possible solutions. These solutions are infinite and function as different shades of a broader colouring that is the problem. To state or pose a problem is a veritable invention, in the sense that the one stating it is not imagining but seeking while fleeing, becoming a 'helpmate to [its] emergence' rather than the originator (Massumi 2009: 40). Invention in relation to experience, the way James and Bergson conceive of it, is happening when a new tone enters a refrain, shifting the manner in which the overall musical landscape was conceived so far. While the old way of tuning into this musical landscape was continuous, a new texture arrived leading to different ways of encountering the musical piece. The past of a certain experience occurs at the same time as the present takes its turn. In a similar way, I would want to pose the problem of politics that resides in both looking for a weapon while looking to drop it in the act of fleeing. The illusion Harney and Moten talk about is an illusion tied to critical thought and economies of critique. Assuming one knew and identified the enemy and thus had the right weapon might be a misleading conception of resistance and self-defence to begin with.

Intuition and invention

Starting with a critique of the notion of common sense, as staged by William James, allows us to emphasise the bifurcation in the modes of thought that Bergson proposes through the concept of intuition. While the critique of Kantian common sense reverberates throughout the works of Deleuze, James' pragmatist hinge allows us to conceptualise intuition against common sense in relation to Bergson's conception of difference. James' pragmatism engages with both dimensions: an epistemological shift towards processes of problematisation replacing knowledge and reason, and an ontological shift that casts every actualised thing as the object of its very own durational subjectivity – thus exploring such formations in their becoming (as well as their repetition) rather than their being. From here a first casting of the method of intuition goes hand in hand with Deleuze's more general critique of the image of thought dominated by the order of the essence, the *what is*, which he opposes with insisting on the minor questions 'Who? How? How much? Where and when? In which case?' (2004: 96).[5] These minor questions are the ones aligned with the fleeing and looking gestures in Moten and Harney. Based on a need to flee, as the historical fact of suppression and violence against delegitimised parts of society, the gestures of looking for a weapon and the need to drop it are instant evaluations – they problematise, based on the overall movement of flight and the movement of singular instances (subjectivities) in light of their need for self-defence and self-preservation. There is a difference in kind between an economy of knowledge and the activity of instantaneous or immanent evaluation. Intuition pertains to such an evaluation as the continuous refrain of a practice that is defined by its genesis, its variation, rather than by its essence. How to think and feel the movement of a problem can neither be answered through knowledge nor through subjective experience. Intuition as a method has to relate to experience's pure state consisting of tendencies and their resonances.

While Bergson's refusal of critical philosophy might seem to remain in an abstract realm, its radicalness as a pragmatist gesture challenges the separation of first and second nature, the given in experience and its ordering

---

5 The text *On the Method of Dramatization* (2004 [1967]) is part of a set of tightly interwoven works starting with *Bergsonism* (1988a [1966]), *Difference and Repetition* (1994 [1968]) and *The Logic of Sense* (1989 [1969]), which form the foundation of Deleuze's notion of becoming.

through abstraction. Politics of a problematising kind do not just account for what is given in experience as always exceeding the conscious grasping. On the contrary, this 'existential grasping' traverses the mental and embodied poles of existence (Guattari 1995: 112). Such a relational understanding ensures that things conceived differentially, along their duration, render matter into an image of the duration in which it inheres, thus foregrounding the ontogenetic character of its very becoming. The auto-affection and 'self-abstraction' (Massumi 2011: 130) inherent in duration as process defines the notion of life in the works of Bergson and Deleuze. It is an utterly impersonal and more-than-human conception of life, a life that is animating and in movement, a creative energetics in the sense that it engages with becoming. Matter is included here as actively moving with and through duration as 'numberless vibrations' (Bergson 1988: 208). Intuition outlines Bergson's humbling proposition to tune into these animating activities of life by accounting for one's own duration 'to affirm immediately to recognize the existence of other durations above and below us' (Deleuze 1988a: 33).[6] To couple life and experience as existential dimensions beyond the organic prepares the ground for a pragmatism based on durational encounters at the core of intuition as a method.

Rather than addressing experience as a sensuous immersion in the here and now, leading towards an abstract order of classification and categorisation through common sense, James foregrounds a pluralist conception of experience as rigorous analytic method. Similar to Bergson's notion of intuition, he insists on making experience not a mere empirical ground from which to abstract in order to obtain generalised notions commonly agreed upon. On the contrary, experience is the only 'stuff' the world is made of, making thoughts and abstractions the same matter as things (James 1996: 4; Bergson 1946: 251). James insists on the mixed states in which human experiences occur. Criticising the objectivist notions of Cosmic Space and Cosmic Time as one Time and one Space he writes: 'The great majority of the human race never uses these notions, but live in plural times and spaces, interpenetrating and *durcheinander*' (1963: 79, emphasis in the original). This *durcheinander* brings forth James' refusal of any separation between first and sec-

---

6 A critical elaboration on two types of vitalist conceptions of life, one of process and one of pathos, in relation to knowledge, problems and their solutions has been developed by Monica Greco (2019).

ond nature that also pertains to Bergson and through which we can understand the very realm for intuition to become active.

Known as the problem of the 'bifurcation of nature' into primary and secondary qualities in Whitehead, Bruno Latour explains the casting of one nature where thought and perception co-emerge:[7]

> If the bifurcation of nature is impossible, then it means that every entity has to explore what, in the rest of the world, may offer it some grasp on life in order for it to continue existing. This grasp is intensely *objective*, since it mobilizes so many other entities; but it is also intensely *subjective*, since it represents, like Leibniz's monads, a very particular version of what the world looks like, that is, an interpretation, a bet, a risk taken, a confidence shared, a choice. (Latour 2005: 234)

The real as developed in James' conception of experience resonates with Latour's Whitheadian take on experience before the bifurcation of nature into primary and secondary qualities. Bergson carves from such a common ground of the real as experience a conception of truth, which conceives of nature as neither a mere given and bearer of facts to be distilled nor an imaginary of the human mind.

'For him [James] those truths it is most important for us to know, are truths which have been felt and experienced before being thought. It has at all times been said that there are truths which have to do with feeling as much as with reason; and that along with those truths we find already made there are also others we assist in the making of, which depend in part on our will.' (Bergson 1946: 253)

In order to access this domain of existence, while not making it a subject of human consciousness, susceptible to good and common sense, Bergson introduces the concept of intuition as method. Following his explorations of experience and the real in James, intuition must be conceived as an affirmative method.[8] Affirmation is not a mere positivism but a way of proceeding by encounter and movement. In affirmation reality provides a 'grip upon it' in moving with it and its flows (1946: 255). Affirming and thus getting a

---

7   On the notion of the 'bifurcation of nature' see Whitehead's *The Concept of Nature* (1920).

8   On the notion of affirmation in philosophy, Deleuze's book on Nietzsche is most informative and highly relatable to Bergson's own use of the term (Deleuze 1983).

grip on reality that 'places us under more favorable conditions for acting' is quite different from knowing in advance how things will play out based on common sense (1946: 255). Tied to experience as the constitutional domain of reality, knowledge follows feeling and intuition is the way to make feeling not a sentiment of the human but a general technique of becoming relationally.[9] 'Intuition starts from movement, posits it, or rather perceives it as reality itself [...] For intuition the essential is change [...] Intuition, bound up to a duration which is growth, perceives in it an uninterrupted continuity of unforeseeable novelty' (1946: 39). This unforeseeable novelty is manifestly bound up with Bergson's understanding of James' conception of truth as not based on what already exists but as bearing a sense of 'what will be' (1946: 255). Intuition then is not a mere mirroring of what nature presents – on the contrary: 'truth, which can be attached only to what we affirm about reality, is [...] created by our affirmation. We invent the truth to utilize reality [...] *While for other doctrines a new truth is a discovery, for pragmatism it is an invention*' (1946: 256, emphasis in the original).

The inventive power of intuition becomes clearest in relation to the staging of problems. The problem of critical theory, as During underlines, is that it cannot account for the problem that does not presuppose a solution. During himself highlights this issue and refers to the positivist concept of problems tied to *problem-solving* rather than *problem-stating* (During 2004: 18). He cites Bergson: 'For a speculative problem is solved as soon as it is well posed', hinting at the inclusion of the solution in a well-posed problem, from which the truth can be uncovered (Bergson in During 2004: 19). In resonance with intuition as the method to not uncover but to invent problems, During refers to the most crucial statement in Bergson's refutation of critiques of problems as uncreative: 'But stating a problem is not simply uncovering, it is inventing [...] Invention gives being to what did not exist; it might never have happened' (1946: 59). In this inventive power of the problem resides the very paradox of intuition and with it the tension between matter and memory or duration. Invention is the term Bergson uses to avoid any relapse into com-

---

9 While there are certainly differences between Bergson's concept of perception and feeling and Whitehead's deployment of the terms, I conceive of feeling as a mode of prehension, the way the term is developed in *Process and Reality* (Whitehead 1987). Therein feeling designates an activity of relational resonance between heterogeneous and varying tendencies entering a process of actualisation.

mon sense. An invention cannot be willed by an individual thought but must be assembled along heterogeneous tendencies and their singular durations – as immanent to experience.

On the one hand, experience occurs before thought, based on feeling, that is, based on a state of being-with before any division into primary and secondary qualities can occur. On the other hand, the stating of problems as inventive acts is tied to affirmation, fabricating a truth to utilise reality. Again, we encounter the logic of the paradox beyond good and common sense. It is clear that Bergson would not diminish his own praise of the pragmatist conception of truth as parsed out through an immediate encounter with the real by conceiving of the future as what can be known through conforming to an established order of knowledge. Truth then, is beyond knowledge, and intuition is tinkering, productive of 'fictions [...] pushing beyond experience a direction from experience itself' (Deleuze 1988a: 25).

Deleuze's very own transcendental empiricist account of a time of experience that itself explodes experience as any given state of affairs is the very time of the event. It is a time smaller than the most minute instant and larger than any conceivable magnitude – what Bergson himself names 'intensive magnitudes' (Bergson 1910: 106). This time outside of any measurable time is the heterogeneous rhythm of durational activity throughout the universe. It is repetition, the very quality of difference as the non-foundational essence of the real. It expresses itself in degrees, in matter, but it can only do so in fleeing, that is, in movement, which contains absolute speeds and slowness but never stillness.

The challenge of the productive paradox of intuition resides in making thought not a faculty of the mind, or rather, to make the mind an aspect of experience. Bergson insists on the co-emergence of a present in its mattering inherence and the past as actively shaping not just the present but its very own tonality as 'memory that prolongs the past in the present' (Deleuze 2004: 28). Conceiving of the past as 'surviving in itself' (as virtual) casts both past and present 'as two extreme degrees coexisting in duration' (2004: 29). For intuition as 'an activity that sets up and organizes problems', this means accounting for things in their difference in duration moving through an alignment to the encompassing duration of which one's own duration is similarly a part of. It further requires us to conceive of thought as immersed in experience and of experience as pushing thought to the boundary of the present by way of accounting for the past's self-sufficiency. In concrete terms this means engaging in a process of becoming, seeking the differential qualities in the encounter of participating and sharing a present at the same time as a past and thus a potential futurity. This futurity, however, lies not beyond the actual present but inhabits its limits. Such is the bounded ecology of experience, a stretching of the limits of the present while acknowledging the multiplicity of tendencies producing and inhabiting such a present. In that sense, as During shows in his account of Bergson's problems, intuition as an activity of stating problems is never outside of history, but its account of the past cannot privilege a commonsensical assumption of what defines *the* past that is relevant for this singular expression of the present.

For the very same reason, Moten and Harney adhere to a different image of thought that cannot operate in the reactive mode of critique but that nonetheless needs to be inventive in terms of drawing on heterogeneous temporalities – this is not an unsolvable knot of eternal complexity, but rather a sobering procedure for effective politics. How so? In the way that the invention of a problem is neither entirely new, that is, ahistorically emerging out of thin air, nor derived from any commonsensical agreement on the past. On the contrary, for problems to preside as political operations, they need to take effect. This hinges on their inventiveness, not as a solely human act but as a co-compositional processing of tendencies in their contribution to an event. Accordingly, 'the mode of the event', in the way Deleuze launches the concept in *The Logic of Sense*, 'is problematic' (1990: 54). As problematic the event adheres to its very own temporality, that of Extra-Being, which operates the real and actualizes partially in the inflexion of a well stated problem.

Tendencies and the politics of the in-act

In his *Essays in Radical Empiricism* James states that 'the experiences of tendencies are sufficient to act upon' (1996: 69) and Deleuze writes, 'what differs in nature is never a thing, but a tendency' (2004: 27). Bergson himself underlines, 'for life is tendency, and the essence of a tendency is to develop in the form of a sheaf, creating, by its very growth, divergent directions among which its impetus is divided' (1910: 99). If intuition as the inventive method of stating problems has to move beyond idealism and realism, it has to confront a world made of tendencies. How, one would ask, can any action based on tendency ever have any purchase in terms of truth? As I have emphasised, the notion of truth requires a radical recasting in light of Bergson, James and Deleuze. It is not just a situated truth but an objectivity that gains relevance because of its singular power to activate the relay between matter and memory as co-emergent. Tendencies are the minimal elements through which the world expresses itself. A tendency is defined by its tending, its movement and its capacity to move in resonance with other tendencies. What needs to be followed through intuition is not 'the presence of characteristics' but the 'tendency to develop' (Deleuze 2004: 34). To know things by their nature means to parse out their very movement of becoming, their tendency. Politically, this means to engage in states of affairs through a 'sense' – which means also direction or tendency in French – of its movement rather than its substantial appearance. Such politics are not ahistorical but rather interlace movements and their sheaves of diverging directions across vast times and territories. Such a tracing, or rather accounting for the heterogeneous emergence of different nuances, is the formation of real problems. The art of such politics resides in the challenge of accounting for new and different nuances that alter the setting, shape the formation and thus provide new perspectives on a concern that seemed to be known.

Coming back to the initial quote, the refusal of critique is paired with a picking up of a weapon while fleeing and dropping it again. The enemy that is an illusion, in a way, is a false problem. How to think about such an abstract coursing concretely, that is, as a problematisation that matters in political practice? Deleuze mentions the term nuance as 'being [which] is the difference itself of [a] thing' as often deployed by Bergson. It actually occurs most notably in *La Pensée et le Mouvant* in the last chapter on Ravaisson and in relation to colour and light (Bergson 1946: 261-300; in the English unfortunately translated as shade). Nuance as differential becomes expressive while always hinting at its 'unifying' ground – which is the past as coterminous with the present (Bergson gives the notion of diffracted light breaking up in colours while still carrying its resonance with white light as the unifying ground). Inserting nuance into the earlier development of intuition, the question of affirming nuances while giving an account of the encompassing problem – an extensive and dynamic complexity – allows us to explore concrete ways of rendering problems into ethical intercessors.

In the takedown of critique, intuition occurs in moving with the situation. The flight is a movement that resonates with a cause but does not resemble it. It senses its quasi-materialisations without reducing it to one or another cause. And further, it re-invents the problem, for instance of suppression, while moving. In the movement it occurs that the cause is in itself an affective field of potential effects – a tendency of its divergent directions. The political activity of intuition concerns the way of engaging with a tendency as it diverges, acknowledging its energetic field while accounting for the occurrent difference in the midst of the present. I would term such politics not the mere act of a volitional subject but an in-act (Manning 2016). It abides by the quest for tendencies as the only *real elements* from which embodied and conceptual effects emerge (see Deleuze 2004: 35). In-acting in a world made of tendencies rather than distributions in time and space affords a specific concept of the subject of action. As much as Harney and Moten refuse the volitional subject of critique, they do not presume that the subject as a social and material confluence of forces is irrelevant. Intuition as a veritable method, however, positions the process of problematisation at the core of any act of creativity. The ontogenetic ground of matter, organisms and thought cannot evolve and endure without an intuition capable of relating tendencies. In-act is the force or an orientation of tendencies towards emergence – it draws on their temporal differences and activates their capacities of resonating with other such differences. Their heterogeneous compositions form the factual outline of embodied and felt experiences, and of thought.

How to become active rather than how to act would be the question I want to raise in relation to politics. Concerning intuition as tied to duration, a pragmatic twist is needed: 'One never commences; one never has a *tabula rasa*; one slips in, enters in the middle [milieu]; one takes up or lays down rhythms' (Deleuze 1988b: 123). Acting is without beginning and end – it becomes a slipping-in rather than a defined act.[10] The notion of the in-act is itself a takedown of action as the political paradigm of a future cause – which would be another variation of the economy that James and Bergson dismissed. Is there a mode of politics that operates through the tending of tendencies and nuances, a speculative-pragmatic practicing in the milieu of process formation? In an untimely fashion, such speculative-pragmatic activations resist common sense logics of what constitutes a problem and how to

---

10 Such an infinitive concept of the act is similar, yet different, to Hannah Arendt's development of the term (Arendt 1958). In another article, I explore the relation between Arendt's conception of the act and Judith Butler's performative take on Arendt's 'spaces of appearance' in detail (Brunner forthcoming).

receive it, while foregrounding the inventive powers of shape-shifting that present intuitively.

The politics of problematisation reside in activating capacitations that intuition is capable of 'inserting' into the unfolding of an event.[11] Politics, or political practice, *qua* intuition, addresses the question of how to engage with the complexities at stake, not undercutting their diverging tendencies while making these differentiating lines apparent beyond foreclosure. However, such an opening of the differential powers of existence is not arbitrary but directed as the inventive threading of a problem. The ethics of the in-act then, address how to insert and relay heterogeneous tendencies in the event's unfolding. It means to engage in the very power of problems as transversal operators capable of activating forms of resistance across various modes of sense and sense-making with their differentiating durations.

In relation to activism, depression and neurodiversity, Erin Manning speaks of the 'art of alignment' (2016: 173), which I see as being in close proximity to the method of intuition. Alignment here is not a submission to an exterior force (in the sense of 'Get in line!'). It means to practice insertion by way of durational resonances and to 'sense' the multiple dimensions of the real capable of co-composing what comes to be expressed as a problem. Manning writes: 'These alignments are not given. They must be crafted. Opening the way for a co-composition that potentially aligns itself to times in the making requires, I believe, a rethinking of the act of alignment itself.' Manning further suggests, *qua* Guattari, that such alignments require the 'account of a collective that exceeds the personal' (2016: 173). This collective is not a group of human subjects – it can be, but more crucially it is the differential quality in tendencies productive of divergent directions. It is also the impersonal that links and courses through divergent tendencies in order to shape a problem and to generate the intersections of matter and memory, past and present. Again, this is not a logic of quantity:

> For the collective as a mode of existence in its own right is not the multiplication of individuals. It is the way the force of a becoming attunes to a transindividuation that is more-than. To become-collective is to align to a chaosmosis in a way that prolongs the capacity of one body to act. (Manning 2016: 173)

---

11 On insertion see Gilbert Simondon (2005: 29).

Problematisation can be viewed as an alignment through encounter with the in-act, as a way of practicing slipping in without a claim to mastery, but with a joy of entering the interplay of durations.[12] In such politics, time matters – it is all about time. It requires the collective activating power at the heart of a problem. Intuition is like a blind-seeing since it does not have a form yet but it very much knows that something is out there. Such is the double nature of fleeing. Writing in the face of the history of violence that takes its roots in the transatlantic slave trade and extends to the deployment of critique as a practice of mastery trained by the whiteness of the Western university, Moten and Harney's hint at fleeing accounts as much for the flight that manifests a genealogy as well as the flight from critique as the redundant return of a hegemonic image of thought. Finally, fleeing is a general movement, a radiation of time that escapes its very capture, it is aion or duration – virtual. The problems of self-defence and self-preservation are well staged in the initial quote. Both are required as forms of maintaining differential lines of existence that resist being subsumed under narratives of common sense or good sense. This is the relentless work of Black Studies to which Harney and Moten refer. But more than that, the looking and dropping of the weapon contains further speculative and pragmatic elements that I conceive as being at the heart of a politics of the in-act.

The in-act is not an act, it is what allows acts to become differentially while aligning to a problem. The problem of institutionalised critique as a continued activity of stating false problems is also a problem of a false conception of the act. It turns the act into an individualised and economic logic, thus rendering it reactive rather than active and affirmative. If the in-act is that continuous ritornello of coming back to a problem's divergent creativity, then the ethics of the in-act is always a collective activation along duration's differential powers. In fleeing, the subject is defined by its mode of traversing, not by its position. In looking for a weapon, a crafting of alignment happens, and in dropping the weapon this alignment passes on into a different situation. The ethical concern or act resides in the differential attunement to the diverging directions, probing them in their shaping of the present, and thus becoming a practice of experimentation. The formation of such *collactives* exceeds the intersubjective scope. Intuition provides a way of tuning

---

12 On a decolonial and feminist critique of modernist narratives of mastery see Julietta Singh's *Unthinking Mastery* (2018).

in to such collactive processes reinventing themselves and becoming uncontainable since they are always in the act and beyond the human (see Manning 2016: 180). The enemy becoming an illusion is not a relativisation of the in-act but of the individualised economy to re-act. Such an affirmative casting of ethics poses challenges to the practice of and need for resistance against powers of capture, violence and extraction. The challenge of the real problem of resistance is one of moving sideways, entering from the middle, expanding the divergent directions of a problem as it meanders and manifests across a variety of past-presents. The logic of the 'counter' – such as counter-powers or counter-effectuation – also requires alternation. It cannot operate by presuming the problem or the enemy in manifest places. It must engage in a plethora of activating flights from capture, in minor gestures, as Manning suggests, and activate their very own durations. The ethics of the in-act as collective process generates relays, resonances and encounters in alliance, that is, with a felt joy of amplification through tendencies. The future then is nothing utopian to adhere to, but in alignment with the in-act coursing through the past-present intersecting in intuition.

References

Arendt, Hannah (1958): The Human Condition, Chicago: University of Chicago Press.
Bergson, Henri (1910): Time and Free Will: An Essay on the Immediate Data of Consciousness, London: George Allen & Unwin Ltd.
Bergson, Henri (1946): The Creative Mind, New York: Philosophical Library.
Bergson, Henri (1988): Matter and Memory. New York: Zone Books.
Bowden, Sean (2018): 'An Anti-Positivist Conception of Problems: Deleuze, Bergson and the French Epistemological Tradition.' In: Angelaki 23/2: pp. 45-63. https://doi.org/10.1080/0969725X.2018.1451461
Bowden, Sean/Bignall, Simone/Patton, Paul (eds.) (2015): Deleuze and Pragmatism, New York: Routledge. https://doi.org/10.4324/9781315764870
Cixous, Hélène (1976): 'The Laugh of the Medusa.' In: Signs: Journal of Women in Culture and Society 4/1: pp. 875-93. https://doi.org/10.1086/493306
Deleuze, Gilles (1983): Nietzsche and Philosophy. European Perspectives, New York: Columbia University Press.
Deleuze, Gilles (1988a): Bergsonism, New York: Zone Books.

Deleuze, Gilles (1988b): Spinoza, Practical Philosophy, San Francisco: City Lights Books.

Deleuze, Gilles (1989): Cinema 2: The Time-Image, Minneapolis: University of Minnesota Press.

Deleuze, Gilles (1990): The Logic of Sense, New York: Columbia University Press.

Deleuze, Gilles (1994): Difference and Repetition, New York: Columbia University Press.

Deleuze, Gilles (2004): Desert Islands and Other Texts, 1953-1974, Los Angeles/Cambridge, MA: Semiotext(e).

Deleuze, Gilles/Guattari, Félix (1987): A Thousand Plateaus: Capitalism and Schizophrenia, Minneapolis: University of Minnesota Press.

Deleuze, Gilles/Guattari, Félix (1994): What Is Philosophy? New York: Columbia University Press.

During, Elie (2004): '"A History of Problems": Bergson and the French Epistemological Tradition.' In: Journal of the British Society for Phenomenology 35/1: 4-23. https://doi.org/10.1080/00071773.2004.11007419

Eshun, Kodwo (2003): 'Further Considerations of Afrofuturism.' In: CR: The New Centennial Review 3/2: 287-302. https://doi.org/10.1353/ncr.2003.0021

Foucault, Michel (1998): Aesthetics, Method, and Epistemology: Essential Works of Foucault 1954-1984, ed. by Paul Rabinow, Vol. 2, New York: New Press.

Greco, Monica (2019): 'Vitalism Now – A Problematic.' In: Theory, Culture & Society, online July 15th. https://doi.org/10.1177/0263276419848034

Guattari, Félix (1995): Chaosmosis: An Ethico-Aesthetic Paradigm, Bloomington: Indiana University Press.

Guattari, Félix (1996): The Guattari Reader, Oxford/Cambridge, MA: Blackwell Publishers.

Harney, Stefano/ Moten, Fred (2013): The Undercommons: Fugitive Planning & Black Study, Wivenhoe: Minor Compositions.

James, William (1963): Pragmatism: And Other Essays, New York: Washington Square Press.

James, William (1996 [1904]): Essays in Radical Empiricism, Lincoln: University of Nebraska Press.

Latour, Bruno (2005): 'What Is Given in Experience?' In: Boundary 32 (1): pp. 223-37. https://doi.org/10.1215/01903659-32-1-223

Madelirieux, Stéphane (2015): 'Pluralism without Pragmatism: Deleuze and the Ambiguities of the French Reception of James.' In: Sean Bowden/Simone Bignall/Paul Patton (eds.), Deleuze and Pragmatism, New York: Routledge, pp. 89-104.

Manning, Erin (2016): The Minor Gesture, Durham, NC: Duke University Press.

Massumi, Brian (2009): '"Technical Mentality" Revisited: Brian Massumi on Gilbert Simondon.' In: Parrhesia 7: pp. 36-45.

Massumi, Brian (2011): Semblance and Event: Activist Philosophy and the Occurrent Arts, Cambridge, MA: MIT Press. https://doi.org/10.7551/mitpress/7681.001.0001

Massumi, Brian (2015): Ontopower: War, Powers, and the State of Perception, Durham, NC: Duke University Press. https://doi.org/10.1215/9780822375197

Quijano, Aníbal (2007): 'Coloniality and modernity/rationality.' In: Cultural Studies 21/2-3: pp. 168-78. https://doi.org/10.1080/09502380601164353

Rifkin, Mark (2017): Beyond Settler Time: Temporal Sovereignty and Indigenous Self-determination, Durham, NC: Duke University Press. https://doi.org/10.1215/9780822373421

Simondon, Gilbert (2005): L'individuation à la lumière des notions de forme et d'information, Grenoble: Millon.

Singh, Julietta (2018): Unthinking Mastery: Dehumanism and Decolonial Entanglements, Durham: Duke University Press. https://doi.org/10.1215/9780822372363

Stengers, Isabelle (2010): 'Experimenting with What is Philosophy?' In: Casper Bruun Jensen/Kjetil Rödje (eds.), Deleuzian Intersections: Science, Technology, Anthropology, New York: Berghahn Books, pp. 39-56.

Stengers, Isabelle (2019): 'Putting Problematization to the Test of Our Present.' In: Theory, Culture & Society, online July 15th. https://doi.org/10.1177/0263276419848061

Whitehead, Alfred North (1920): The Concept of Nature, Cambridge, UK: Cambridge University Press.

Whitehead, Alfred North (1987): Process and Reality: An Essay in Cosmology. Gifford Lectures Delivered in the University of Edinburgh During the Session 1927-28, New York: Free Press.

Pragmatics of a World To-Be-Made

Martin Savransky

Paradoxes

We might as well begin with a paradox. After all, how is the problematic expressed if not through a sort of paradox of our present, one whereby the present becomes fugitive, boiling over itself, constituting a time 'while passing in the time constituted' (Deleuze 1994: 79)? So does the proposition of this book, of *thinking the problematic*, confront us with a paradox in which the problematic makes itself manifest, from which it cracks open, proffering itself fugitively in search of new presents. And the paradox is this: What does thought ever do, if it does not think the problematic? What is thinking if not the event of becoming possessed by a problematic one cannot shake, let alone properly state, a problematic that spurs the thinker into thinking, feeling and doing? This is what William James (1890: 401) alluded to when he suggested that 'the thought is itself the thinker, and psychology needs not look beyond'. For the thinker is constituted as such by a problematic for which it becomes a *means*.[1] James expounded on this idea with his concept of a 'fringe', a fringe of felt relations on the edge of which thoughts – which is also to say, thinkers – swim. The fringe constitutes a vector of indetermination, and in 'all voluntary thinking there is some topic or subject about which all the members of the thought revolve. *Half the time this topic is a problem*, a gap we cannot yet fill with a definite picture, word, or phrase, but which, in the manner described some time back, influences us in an intensely active and determinate psychic way. Whatever may be the images and phrases that pass before us, we feel their relation to this aching gap. *To fill it up is our thought's destiny.*' (James 1890: 80)

---

1 I am thankful to Isabelle Stengers (2014) for this expression.

If the problematic acts as the generative force, the paradoxical imperative of adventure of which the thinker becomes not its hero but its path, what kind of gesture, which sort of operation, might be at stake in *thinking the problematic*? Which is to say, to what kind of adventure are we propelled to when we ask of thought to fold onto itself, to *complicate itself* in order to think that which makes it think? What difference might this complication make? To which new paradoxes might it give rise to? And what new possibles may such paradoxes crack open? Of course, learning to appreciate this generative recursion of paradoxes requires, in the first instance, that we consent to a radical reversal concerning the problematic itself. This is the radical reversal to which Gilles Deleuze (1994: 158) submitted the very notion of a 'problem', when he sought to dissociate it from that 'grotesque image of culture' which infects and glosses over, with equal force, both the constitution of our present and the very mode of passing of the present in the time it constitutes. A grotesque image of culture that has been at the heart of modern colonialism and global capitalism, and has infiltrated modern state politics and development programs, environmental policy and global health, but also 'examinations and government referenda' as well as 'newspaper competitions (where everyone is called upon to choose according to his or her taste, on condition that this taste coincides with everyone else)' (Deleuze 1994: 158).

This is the image that turns the problematic into an obstacle to be overcome, and renders problems mere shadows of their eventual solutions. Under such an image there is indeed no apparent paradox involved in thinking the problematic. The proposition becomes equivalent with *solving* problems. This is because, according to this image, problems are not just given – they are given ready-made. All that matters is to find the right solution, the one that will eventually make the problem a mere figment of the unlearned world, an irrelevance, an innocent vestige of our past ignorance. Even as problems become 'wicked', 'fuzzy', or 'complex', the sense of the problematic that our culture espouses is one that treats it exclusively as an epistemic puzzle – an obstacle posed to our knowledge, to our methods; a matter for thought and science alone (followed by the acknowledgment that the more sciences involved, the better in driving the problem to its own exhaustion). Here, thinking becomes an act of exhaustion of problems in solutions for which thoughts and sciences are never their *means* but their masters. And indeed, just as the problems are given ready-made, so are the solutions. They may not be apparent to the ignoramus who is confronted by the problem that a teacher sets out in an examination, or to the state that is confronted with a response to a referendum on an ill-posed question it did not even understand, but the very staging of the problem presupposes that a solution *must* exist, that it is a matter of picking the problem apart so as to find it, a matter of identifying the solution with a truth that the problematic itself occluded – *the people have spoken*. It is always, in the end, a matter of puzzle-solving. The image of the completed puzzle is printed on the box which contains it – all one needs to do is to copy, to imitate, to find the corner pieces that already determine the contours of the puzzle or problem and simultaneously enable the derivation of the only true solution, the one that reproduces an image that is identical to the one given at the very outset. Indeed, that is what this grotesque image of culture turns the problematic into – a puzzle, a veil, a blockage, a temporary obscurity, a shadow of knowledge, an obstacle to be overcome by following the right example, by deploying the appropriate methods.

Deleuze's gesture would then consist, in the first instance, in noting that inside and in spite of this grotesque image of culture the paradoxes persist and insist. The puzzle is never finished, and the solutions the moderns come up with never quite exhaust the problematic. This is why he associates such image of culture with the notion of 'stupidity', *la bêtise*, characterizing it after Bergson as nothing other than a 'faculty for false problems', the 'evidence of an inability to constitute, comprehend or determine a problem as such' (Deleuze 1994: 159; see also Debaise 2016). Because this culture of puzzle-solving, which is our own, cannot but continue failing to accomplish that which it sets out to achieve – the complete exhaustion of the problematic as such, the dream of a universally valid Reason, of a perfectly frictionless world, the perpetual peace of a permanently smooth present. The paradox is of course that the problematic presses on, 'it insists and persists in these solutions' (Deleuze 1994: 163) such that the latter do not ever solve problems without also making them proliferate in new ways, provoking new imperatives to which thought is forced to respond. This is why it 'would be naïve to think that the problems of life and death, of love and the differences between the sexes, are amenable to their scientific solutions and positings, even though such positings and solutions necessarily arise without warning, even though they must necessarily emerge at a certain moment in the unfolding process of the development of these problems.' (Deleuze 1994: 107)

If paradoxes constitute sites where the problematic cracks open, from which it creates a line of fugitivity, the second aspect of Deleuze's gesture is precisely to trace this line, to follow the problematic outside of this image of culture that has sought, and failed, to contain it. And it is there, outside, that the problematic can no longer designate a mere state of ignorance or imperfection, for it can never be contained in the knowledges that would seek to dissipate it in their solutions. Outside of this grotesque image of culture, the problematic becomes 'a state of the world, a dimension of the system, and even its horizon or its home' (Deleuze 1994: 280) – an occasion of experience boiling over onto a new occasion, the thought streaming through the thinker it has brought into being, the present passing in the time it has constituted, the world opened up to its own becoming. 'Let anyone try', James (1890: 608) wrote prefiguring this gesture, 'I will not say to arrest, but to notice or attend to, the *present* moment of time. One of the most baffling experiences occurs. Where is it, this present? It has melted in our grasp, fled were we could touch it, gone in the instant of becoming.'

In this other culture of paradoxes, this culture without image that James's and Deleuze's gestures help us conjure, the problematic can no longer merely correspond to a shadow of knowledge, for it is the present itself that crumbles in our grasp. How to characterise this crumbling present, which is also the calling forth of the present moment by the insistence of another present that urges the fugue? In what sense may the problematic constitute, in passing, the *home* and *horizon* of the world if it does not designate a specific *mode of existence, the generative mode of existence of a world to-be-made*? It is this version of the problematic that I am seeking to *think*, or rather, to try and explore some of what might be at stake in our thinking it. This version in which the term 'problematic' conjures, with a word, the lure of the world's own fringes, the sirens of what is in the process of being brought in, of a buzzing possibility, of a difference to come (Savransky 2018a). To actualise it is the world's destiny. But here's another paradox: such destiny is never guaranteed. And so we may propose, as a working hypothesis that relays the paradox we started with, that perhaps the task involved in *thinking the problematic* is no other than a gesture of learning, experimentally, how to relate to the fringe, how to sustain and dramatise the process through which a possible makes its insistence felt with the character of an imperative, by which the an insistent possibility irrupts and reconfigures the world made. The task might be, in other words, that of developing a pragmatics of a world to-be-made.

Metamorphoses

An attentive reader may have noted an air of familiarity in this notion I have just associated the problematic with, that of the mode of existence of a world to-be-made. It is, of course, borrowed, in homage and in relay, from the expression used by another philosopher, Étienne Souriau (2015), in his lecture titled 'Of the Mode of Existence of the Work To-Be-Made'. And such borrowing is quite deliberate. For if Deleuze's gesture enables us to trace the problematic along its fugitive lines, outside the false problems of our puzzle-solving culture, it seems to me that Souriau's essay dramatises with unique taste and ability the task before us here – that of learning how to characterise that process by which a possible makes its insistence felt with the character of an imperative. In a sense, Souriau's (2015: 220) *problem* is of course quite different – 'Is existence ever a piece of property that we possess? Is it not rather an objective and a hope?' One might hasten to see this as a mere permutation of the perennial problem we call 'ontology': What does it mean for something 'to exist?' And it is that, but *not only*.[2] For once again, under the auspices of our puzzle-solving culture, we have treated most philosophical ontologies as so many *solutions* to this problem. With Souriau, by contrast, the resonances are made possible not least by the fact that he is concerned, above all, with ensuring 'that [his] problem is well-posed': how to think the problem of characterizing something as existing?

Attempting to pose the problem anew, to think the problematic that forces him to think, Souriau (2015: 220) experiments with a dramatic hypothesis: that the problem of existence may not involve a binary choice, but may after all be better approached as a problem of *intensity*, such that, 'in response to the question, "Does that being exist?" it is prudent to admit that we can hardly respond with the Yes-No couple, and that we must instead respond in accordance with that of the More or Less'. Once 'being' ceases to be a question of 'yes or no' and becomes a matter of intensity and degree, the entire sense of the problem of ontology changes, for it becomes a question of a plurality of modes of existence, of the varying degrees of (in)completion of things, and crucially, of the pragmatic question of their *genesis* – that is, of the creative accomplishment of their existence. It is this process of creative accomplishment – of *instauration*, as he calls it– that makes it possible to ask the question

---

2 For a very generative use of the 'not only' see Marisol de la Cadena (2014).

of the mode of existence, and therefore also of the generative force, of that which is still in the making, of 'the work to-be-made'. It is important to note, however, in which sense Souriau conceives of this generation of existence that implicates all *relatively* existing things (for indeed, when existence is a matter of intensity, one only exists *relatively*): 'We all know', he writes, 'that each of us is the sketch of a better, more beautiful, more grand, more intense, and more accomplished being, which, however, is itself Being to-be-realised, *and is itself responsible for that realization*' (220). In other words, when existence is a matter of intensity, a possible still in the making must nevertheless have some dim existence of its own, an existence whose generation is neither a case of spontaneous 'self-realization' nor one of the wilful 'construction' of one being by another. As such, just like a thinker is brought into thinking as it becomes the means of responding to a problem that makes her think, 'the accomplished existence, here, is not only a hope, but also responds to a power' (Souriau 2015: 220).

Indeed, I would like to suggest here that the problematic may have something of this character too, of a yet-unmade world that nevertheless makes itself felt with imperative force, that 'imposes itself as an existential urgency – which is to say: both as deficiency and as presence of a being to be accomplished, and which manifests itself as such, as having a claim on us' (Souriau 2015: 223). But is this not just a spurious analogy? Is Souriau not dealing, after all, with an altogether different problem? I don't think so, not entirely.[3] And the reason for this is that, if when ontology is treated as a binary problem (this exists, this does not) the question is where to draw the line, when it becomes a problem of intensity and degree the question is how to think the *intensification* of existence. Which is also to say, how to characterise one's relationship to the fringe – the relationship of the thinker to the thought for which it is in process of becoming a means, of the constituted present to the one that is passing in the time constituted, of the world made to the world to-be-made. Not unlike ours, Souriau's problem is, in other words, a problem of heterogenesis, of the *actualisation of a possible*, of the determination

---

3 The resonances are also not entirely coincidental either – Deleuze and Guattari (1994: 220, n, 6) indeed acknowledged their debt to Souriau in *What is Philosophy?*, and as Isabelle Stengers and Bruno Latour (2015: 13) note in their introductory essay to Souriau's *The Different Modes of Existence*, there are already hidden references to Souriau in Deleuze's *Difference and Repetition* —references to the work of art to-be-made, and to the virtual as a task to be performed— that are 'as plain to see as the famous purloined letter of Edgar Allan Poe'.

a problem, of the generation of a being that is 'only able to be accomplished completely through the power of another being' (Souriau 2015: 223).

It is in order to dramatise this process that Souriau (ibid: 225) pays attention to the very activity of making, and provides us with a most dramatic account of the process of sculpting:

> Watching the work of the sculptor, I see how with each blow of the mallet and chisel, the statue, at first a work to-be-made, absolutely distinct from the block of marble, is gradually incarnated in that very marble. Little by little, the virtual work is transformed into a real work. Each of the sculptor's actions, each blow of the chisel on the stone constitutes the mobile demarcation of the gradual passage from one mode of existence to another.

At stake in this process of sculpting is not, therefore, a mere act of 'human creativity' or of 'imagination', a simple process of projection, of the impressing of a human will on an amorphous thing by means of the chipping away of the marble. To put it more bluntly, it would be entirely wrong even to suggest that the marble, the sculptor, and his instruments constitute the only characters in play. Of course the statue will not be made by itself, and neither 'will future humanity. The soul of a new society', Souriau (2015: 227-228) wrote, 'is not made by itself, it must be worked toward and those who work toward it really effect its genesis. [...] If our sculptor – weary, having lost faith in his work, incapable of resolving the artistic problems that stand between him and the possibility of advancing – lets the chisel fall or stops striking it with the mallet, the work to-be-made remains in limbo.' Nevertheless, the statue is present too, from the very outset, as a work-to-be-made, as a generative *problem* that turns the sculptor into its means. It is sculpting's own *destiny*, yet it is never guaranteed. What's more, its dim existence is highly demanding, a veritable test, pressing on the sculptor not with ready-made gestures that the latter may simply apply on the marble, but with 'the ever recurring questions of the sphinx: 'work it out, or thou shalt be devoured.' But it is the work that blossoms or vanishes, the work that progresses or is devoured.' And yet, the work remaining in limbo is not the only risk that such heterogenetic process is faced with. Heterogenesis is a thoroughly experimental process, and one can only proceed piecemeal, 'groping our way forwards like someone climbing a mountain at night, always unsure if his foot is about to encounter an abyss' (ibid: 229).

In other words, this statue to-be-made, this being-of-the-fringe, constitutes a real character in the process of its own heterogenetic intensification. Tempted yet reluctant to conceive of it as a 'person', Souriau decided to call this character 'the Angel of the work'. But the experiment of sculpting the statue may well fail to respond to the Angel of the work, to the statue to-be-made, thereby leaving the sculptor frustrated: just as those of us who write may feel the sense of frustration at the accomplished reality of a text that, when on the page, is not what it could have been; just as we feel a sense of diminishment when the words we utter in a conversation seem unworthy of the idea that we are trying to conjure. This is because the Angel of the work does not constitute an *answer* to our problem, one that would be given ready-made. Unlike the experience of being unable to solve a simple problem of arithmetic, the frustration that comes from the failure of an experiment in intensification is not one that reveals our 'ignorance', but the feeling of a certain devaluation, a poverty, a barrenness, of that which has been made actual. And such a feeling makes present that, rather than designate an answer to our problem, the Angel of the work constitutes *the very problematic* to which we seek to respond, establishing with us a *'questioning situation'* that demands a response but does not dictate what that response shall be. As Souriau (2015: 232) puts it, ultimately unable to shake the temptation to characterise the Angel as a person, the work to-be-made never says '"Here is what I am, here is what I should be, a model you have only to copy." Rather, it is a mute dialogue in which the work seems enigmatically, almost ironically, to say: "And what are you going to do now? With what actions are you going to promote or deteriorate me?"'

Which is also to say that, insofar as the problematic demands a response, insofar as it makes itself felt with existential urgency but does not say what the correct answer will be, every intensification of its existence involves a process of *metamorphosis*. That is, at one and the same time, a heterogenetic transformation: of the world made, whose actuality progressively becomes torn at the seams by the demanding insistence of a world to-be-made; and of the problematic itself, transformed in its being drawn in, in its concrete intensification as a member of *this* world, in its progressive development into a specific problem and its associated field of solvability – always necessary, always insufficient, 'for in every realization, whatever it may be, there is always a measure of failure' (ibid: 236). Indeed, I would suggest this is how we could read Deleuze's (1994: 107) own remark, that it 'may be that there is something mad in every question and every problem, as there is in their transcendence in relation to answers, in their insistence through solutions and the manner in which they maintain their own openness.' Because the generative force of the problematic always comes from the fringe, from an otherwise, another world in this world, from somewhere else than its determinations into propositions and solutions. As Arundhati Roy (2005: 44) once proposed to the World Social Forum: 'Another world is not only possible, she's on her way. Maybe many of us won't be here to greet her, but on a quiet day, if I listen very carefully, I can hear her breathing.'

Conjurings

Coming from another world in this world, the insistence of a world to-be-made cannot be satisfied by reasons capable of explaining *why* a problem has presented itself with such intensity, turning one into its very means of intensification. Persisting, after every attempt to respond, after every gesture of intensification, with its nagging question, 'And what are you going to do now? With what actions are you going to promote or deteriorate me?', the problematic acts as a vector of generativity introducing an after to every ending. This is what plural and collective movements trust, those who, in various non-colonial languages like Quechua, Guaraní, or Urdu, experiment with a plurality of efforts and calls to protect and intensify our relation to food, land, water, Pachamama, dignity, or *buen vivir* (e.g. Fisher and Ponniah 2015). What they trust is that this other possible world which they seek to intensify is not *only* a hope or the endpoint of a project, but a world underway that insists with existential urgency, that makes a claim upon them. In other words, their calls too are responses to a power, they are attempts to induce a metamorphosis that, in laying siege to the imperial, corporate force of what has come to be known as 'globalization', might become capable of intensifying a multiplicity of other worlds to-be-made. And if their efforts and claims sound 'mad' or 'naïve' to the modern ears that hear them, this is because, whatever their fate, they already begin to rip the very culture of 'puzzle-solving' at its seams. It is, in other words, because their calls and efforts already make present that, to borrow Deleuze's (1994: 158) words again, we risk remaining 'slaves so long as we do not control the problems themselves, so long as we do not possess a right to the problems, to a participation in and management of the problems'.

We have to stress this point; the openness of every question and problem would constitute a form of 'madness' only within that grotesque image of culture which makes the very task of 'thinking the problematic' something of a non-starter. For indeed, if as Souriau (1948: 226) suggested once, 'culture is a style of thinking and doing that guides, towards a certain form of feeling, everything that is mobilised and elaborated by the instaurating forces of a human group', it may well be that, in reducing the problematic to a mere state of ignorance, it is the puzzle-solving style of our culture that cannot but confuse the feeling of existential urgency of a world to-be-made with a kind of *disorder*. As Thomas Kuhn (2012) said of the periods of 'normal science' which he, not innocently, characterised as fundamentally concerned with 'puzzle-solving':

> Perhaps the most striking feature of normal research problems [...] is how little they aim to produce major novelties, conceptual or phenomenal. Sometimes, as in a wave-length measurement, everything but the most esoteric detail of the result is known in advance, and the typical latitude of expectation is only somewhat wider. Coulomb's measurements need not, perhaps, have fitted an inverse square law; the men who worked on heating by compression were often prepared for any one of several results. Yet even in cases like these the range of anticipated, and thus of assimilable, results is always small compared with the range that imagination can conceive. And the project whose outcome does not fall in that narrower range is usually just a research failure, one which reflects not on nature but on the scientist. (2012: 35)

This may be striking, or it may not. For if it is this style of puzzle-solving that also leads us 'to believe that the activity of thinking, along with truth and falsehood in relation to that activity, begins only with the search for solutions, that both of these concern only solutions' (Deleuze 1994: 158), then the very possibility of a problem that insists and persists in its solutions, a problematic that demands to be thought, one whose demands often burst into the world in the form of completely unexpected results, would become a sign not of novelty but of error. Because the gesture of responding to the problem of another possible world, of intensifying a world to-be-made, can no longer be a matter of coming up with a truth capable of making the problem disappear. By contrasts, it involves the development of a veritable art of *conjuring* the being-of-the-fringe, of learning how to attend to the demanding questions it poses.

Not coincidentally, this is precisely what other styles of thinking and doing, those 'cultures' that – as is for instance the case in some corners of contemporary Cuba – live with both, scientists and oracles, are sometimes aware of (see Holbraad 2012). Because of their self-cultivation as artful conjurers, oracles of the Cuban Ifá tradition cannot *but* speak the truth – indeed, they cannot *but* speak a form of truth that our culture of puzzle-solving has banished as a chimera, namely, a truth that is fundamentally *indubitable*. The oracle's practice of veridiction consists precisely in conjuring a response to the consultant's concern, one that becomes intensified as they, in their doings, progressively bring together different, dynamic paths of existence and meaning – the mythical path of Ifá gods, the meaning emerging of the manipulation of the material powders and paraphernalia used during the consultation, and the personal path of the consultant – such that a metamorphosis of all such trajectories can be accomplished (Holbraad 2012).

And yet, when the oracle speaks the truth, the verdict is often bewildering to those that consult them.[4] The truth is itself a problem to which the consultant must invent a response, inducing yet another metamorphosis – of their own life, affected by the verdict of the oracle, and of the problem the verdict has posed, eventually actualised in the situated actions that the consultant takes in relay and return. The test, in any case, is what *kind of transformation* the oracle's problem gives rise to. For if the problematic acts as a generative vector, as a demand for intensification of a possible, a call that lures the world and one's life to keep on going differently at the fringe, the challenge facing any solution is not whether it is true or false, but whether, with its response, it promotes or deteriorates the intensity of the possible that insists at the edge of the present. If solutions there will be, the task is to ask of them *how* they might make the world go on. The pragmatic test insists again: What difference will they make?

---

4 Interestingly, Deleuze (1994: 63) made a very similar point in relation to the oracles of ancient Greece: 'Myth tells us that it [a grounding] always involves a further task to be performed, an enigma to be resolved. The oracle is questioned, but the oracle's response is itself a problem.'

To dissociate the notion of solutions from the dream of truth, to recognise that there 'are no ultimate or original responses or solutions' (Deleuze 1994: 107), does not, then, lead to a free-for-all attitude, a 'whatever works'. The test of truth disappears from the nature of solutions only to multiply itself on other levels. For if a 'solution always has the truth it deserves according to the problem to which it is a response', truth and falsehood are engendered in the problematic itself, such that it 'has the solution it deserves in proportion to its own truth or falsity – in other words, in proportion to its sense' (Deleuze 1994: 159). Is this problem genuine? Does it effectively present itself with existential urgency, making a claim on us, leaving no standing place outside of the alternatives it creates? This is why, when the oracle's verdict is too far removed from anything that may enable consultants to feel its presence with intensity, the concerns the latter may develop are not whether the verdict is actually *true*, but whether the oracle has conjured a genuine problem – in other words, whether the one conjuring it is in fact an oracle (Holbraad). In this way, what displacing the genesis of truth to the problematic makes possible is a metamorphosis of the very relationship between problems and solutions, harnessing the irrepressible generativity of problems and questions while submitting solutions to a pragmatic challenge. If the best that a solution can do is to *develop* a problematic, to promote its existential intensity, what is required is a participation in the conjuring of problems themselves. Which is to say, an experimental cultivation of the arts and operations of conjuring that a problematic may require for the vectorizing of a metamorphosis – one that redraws the contours of what a generative formulation of the problem might be, and what it may demand of us on this day that, if successful, will no longer be 'today'.

Presentiments

What is at stake in this pragmatic metamorphosis of problems and solutions, then, is a different kind of responsiveness to the problematics that make us think, feel, and do – a kind of responsiveness that might situate a multiplicity of divergent practices and collectives in the face of a shared perplexity, articulating responses that comprehend and appreciate without demanding salvation, responses that can refuse participation in settled modes of problematisation without their refusal coinciding with a cynical dismissal of the reality of the problematic as such (Savransky 2018b). And what this pragmatic metamorphosis perhaps enables, in turn, is the elaboration of responses whose task is neither to be 'right', nor to achieve a definition of a problem that no one could refuse; but responses, instead, that may seek to collectively experiment with the imperative that the problematic itself creates at the age of the present – practices capable of conjuring, intensifying, and consenting to the metamorphic process of generating and responding to worlds to-be-made. And here we can see another multiplication of the question of truth: for the test of those practices involved in the generation of such responses will not be a test of adequacy, controlling whether, with their solutions, their intensification corresponds to the state of affairs of the world made. By contrast, it will be one of *verification* in the pragmatic sense, that is, of their eventual success or failure in *effecting* such metamorphosis, in *making* a transformation of our world *true*.

William James (1988: 237) once said that the distinct mark of pragmatism is, precisely, that whereas other philosophies postulate a pre-existent and absolute truth that our ideas must imitate, 'the pragmatist postulates a "reality" for our ideas to be become true of.' A pragmatics of a world to-be-made, after all: the crafting of a response, to the tearing at the seams of our present, by the intensification of a fugitive present that passes in the time it has constituted. Indeed, 'if those who think about a future world to be made to come into being did not, in their dreams of it, find some wonderful *presentiment* of the presence for which they call, if, in a word, the wait for the work was amorphous, there would doubtlessly be no creation.' (Souriau 2015: 230) Which is also to say that, if learning to cultivate generative and heterogeneous relations to those beings-of-the-fringe involves consenting to a pragmatic metamorphosis of the passing, into one another, of our world made and a world to-be-made, then this consent can never be a matter of 'thought' in abstraction from the feeling of a fringe that this fugitive present, this yet-unmade world calling the world made forth, makes felt with the character of an imperative.

Thus, whenever it is a matter of thinking the problematic, thought can never become a well of originary gestation, but is always a vector of transformation of a problematic field – the gesture, at the edge of the present, of dramatizing the feeling of the fringe, of enabling the passing of another world-in-this-world to become a vector of thought. And in this sense, it is entirely apposite, it seems to me, that Souriau would call this feeling a 'presentiment'

– for here the prefix is not attached to the sentient experience itself, but to the dim existence of that which *makes us* feel. To call it a presentiment is to emphasise that this sentience corresponds to the feeling of an 'if' rather than an accomplished 'is'. It is the feeling of a possible that demands to be honored, that calls for its own intensification. To think the problematic, then, may well amount, quite simply, to *trusting* those presentiments. It may amount to giving to the 'if' that makes us feel the tools it may need, so that, at the edge of a present that wonders *how* to go on, it may paradoxically introduce, in the world made, the difference required for the invention, always at risk, always unfinished, of a different sense – of another world to-be-made.

References

Bergson, Henri (2007): The Creative Mind: An Introduction to Metaphysics, Mineola: Dover.

De la Cadena, Marisol (2014): 'Runa: Human but not only.' In: Hau: Journal of Ethnographic Theory 4, pp. 253-259. https://doi.org/10.14318/hau4.2.013

Debaise, Didier (2016): 'The Celebration of False Problems.' In: Bruno Latour (ed.), Reset Modernity!, Cambridge, MA: MIT Press.

Deleuze, Gilles (1994): Difference and Repetition, New York: Columbia University Press.

Deleuze, Gilles/Guattari, Félix (1994): What is Philosophy? New York: Verso.

Fisher, William F/Ponniah, Thomas (2015): Another World is Possible: World Social Forum Proposals for An Alternative Globalization, London: Zed Books.

Holbraad, Martin (2012): Truth in Motion: The Recursive Anthropology of Cuban Divination, Chicago: University of Chicago Press. https://doi.org/10.7208/chicago/9780226349220.001.0001

James, William (1890): The Principles of Psychology, Vol. 1, Mineola: Dover. https://doi.org/10.1037/10538-000

James, William (1988): Manuscripts, Essays, and Notes, Cambridge, MA: Harvard University Press.

Kuhn, Thomas (2012): The Structure of Scientific Revolutions, Chicago: University of Chicago Press.

Roy, Arundhati (2005): An Ordinary Person's Guide to Empire, New Delhi: Penguin.

Savransky, Martin (2018a): 'The Humor of the Problematic: Thinking with Stengers.' In: SubStance 47, pp. 29-46.

Savransky, Martin (2018b): 'The Social and Its Problems: On Problematic Sociology.' In: Noortje Marres/Michael Guggenheim/Alex Wilkie (eds.), Inventing The Social, London: Mattering Press.

Souriau, Étienne (1948): 'La culture et les respecte des cultures.' In: Études philosophiques, June, pp. 226-230.

Souriau, Étienne (2015): The Different Modes of Existence, Minneapolis: Univocal.

Stengers, Isabelle (2014): 'Speculative Philosophy and the Art of Dramatization.' In: Roland Faber/Andrew Goffey (eds.), The Allure of Things: Process and Objects in Contemporary Philosophy, London: Bloomsbury.

Stengers, Isabelle/Latour, Bruno (2015): 'The Sphinx of The Work.' In: The Different Modes of Existence, Minneapolis: Univocal.

About the Authors

Christoph Brunner is Juniorprofessor for Cultural Theory at Leuphana University Lüneburg. In his work he deals with the politics of affect in activist and aesthetic media practices in translocal and transtemporal contexts. He initiated the *ArchipelagoLab for Transversal Practices*. In 2019 he received the John G. Diefenbaker Award for a one year visiting professorship at McGill University Montreal with the book-project 'Activist Sense: Towards an Aesthetic Politics of Experience'. His works have been published in Conjunctions, Third Text, transversal, Open!, Journal for Aesthetics & Culture amongst others.

Thomas Ebke (Dr. phil.) is currently academic assistant at the chair for political philosophy and philosophical anthropology at the University of Potsdam, Germany. His fields of research comprise philosophical anthropology, the tradition 'historical epistemology' (particularly Canguilhem) and 20th century French philosophy. His ongoing book project is the elaboration of a systematic 'metaphysics of difference' based on Jean Hyppolite's reception of Hegel's philosophy. Selected publication: 'Lebendiges Wissen des Lebens. Zur Verschränkung von Plessners Philosophischer Anthropologie und Canguilhems Historischer Epistemologie' (Akademie Verlag, 2012; 'Vital Knowledge of Life. On the entanglement between Plessner's philosophical anthropology and Canguilhem's historical epistemology').

Oliver Leistert (Dr. phil.) is a media and technology researcher at Leuphana University interested in the effects of media-technologies on subjectivation processes and sociality, and in philosophies of technology. He is principal investigator for a DFG-funded research project on the environmentality of blockchains entitled *Blockchains. Media of Sovereignty*. Recent publications include 'Governing Objects from a Distance: Blockchains as Organizers of

Environmentality', in: Burckhard et al. (eds.): Explorations in Digital Cultures (Meson 2020); with Lina Dencik (eds.): Critical Perspectives on Social Media and Protest. Between Control and Emancipation (Rowman & Littlefield, 2015).

Celia Lury is a Professor in the Centre for Interdisciplinary Methodologies, Warwick University. Her three main areas of research interest are: feminist theory and the category of gender, sociology of culture, and interdisciplinary methodologies. She has recently co-edited the 'Routledge International Handbook of Interdisciplinary Research Methods' (2018).

Esther Meyer, is a Doctoral Candidate at the Faculty of Sustainability, Leuphana University. Currently, Transformative Research Project Manager at Lighthouse, working on social-ecological sustainability action and research with civil society organizations; recently co/published: 'Solvable problems or problematic solvability? Problem conceptualization in transdisciplinary sustainability research and a possible epistemological contribution', in: *Gaia* 29/1, 2020; 'Designing a Transformative Epistemology of the Problematic: A Perspective for Transdisciplinary Sustainability Research', in: *Social Epistemology* 34, 2020.

Martin Savransky is Senior Lecturer in the Department Sociology at Goldsmiths, University of London. He is the author of 'Around the Day in Eighty Worlds: Politics of the Pluriverse' (Duke University Press, 2021) and 'The Adventure of Relevance' (Palgrave, 2016). He is co-editor of 'Speculative Research: The Lure of Possible Futures' (Routledge, 2017), guest-editor of 'Isabelle Stengers and the Dramatization of Philosophy' (SubStance, 2018) and 'Problematising the Problematic', in: *Theory, Culture & Society*, 2021. He has published widely across philosophy, cultural studies, postcolonial studies and the social sciences.

Isabell Schrickel is a PhD candidate at the Center for Global Sustainability and Cultural Transformation (Leuphana/Arizona State University) where she finishes her dissertation on the history of the *International Institute for Applied Systems Analysis* IIASA. Isabell was a visiting fellow at the Harvard Department of the History of Science. Her research interests include the history of the environmental sciences, the history of modeling and simulation, and the evolution of sustainability thinking. Recent publications: 'Decision Support System', in: Krajewski et al. (eds.): Enzyklopädie der Genauigkeit (Diaphanes, 2020); 'Control versus Complexity: Approaches to the Carbon Dioxide Problem at IIASA', in: *Berichte zur Wissenschaftsgeschichte* 3, 2017.

Jean-Baptiste Vuillerod is Doctor of philosophy. He wrote a PhD on the reception of Hegel's philosophy in France, especially in Louis Althusser's, Michel Foucault's and Gilles Deleuze's work. He edited Jacques Martin's manuscript on 'L'individu chez Hegel' (ENS éditions, 2020) and a collective volume entitled 'Adorno contre son temps' (Presses universitaires de Nanterre, 2019). He also works on the relationship between Hegelianism and Feminism in his book 'Hegel féministe. Les aventures d'Antigone' (Vrin, 2020).

Social Sciences

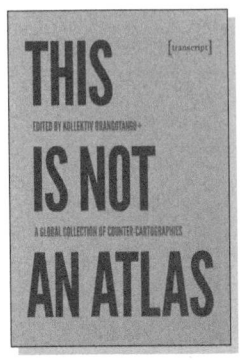

kollektiv orangotango+ (ed.)
This Is Not an Atlas
A Global Collection of Counter-Cartographies

2018, 352 p., hardcover, col. ill.
34,99 € (DE), 978-3-8376-4519-4
E-Book: free available, ISBN 978-3-8394-4519-8

Gabriele Dietze, Julia Roth (eds.)
Right-Wing Populism and Gender
European Perspectives and Beyond

April 2020, 286 p., pb., ill.
35,00 € (DE), 978-3-8376-4980-2
E-Book: 34,99 € (DE), ISBN 978-3-8394-4980-6

Mozilla Foundation
Internet Health Report 2019
2019, 118 p., pb., ill.
19,99 € (DE), 978-3-8376-4946-8
E-Book: free available, ISBN 978-3-8394-4946-2

**All print, e-book and open access versions of the titles in our list
are available in our online shop www.transcript-verlag.de/en!**

Social Sciences

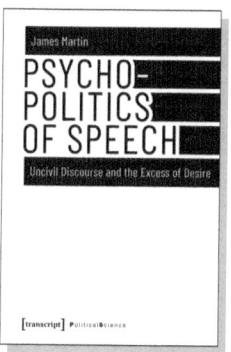

James Martin
Psychopolitics of Speech
Uncivil Discourse and the Excess of Desire

2019, 186 p., hardcover
79,99 € (DE), 978-3-8376-3919-3
E-Book: 79,99 € (DE), ISBN 978-3-8394-3919-7

Michael Bray
Powers of the Mind
Mental and Manual Labor
in the Contemporary Political Crisis

2019, 208 p., hardcover
99,99 € (DE), 978-3-8376-4147-9
E-Book: 99,99 € (DE), ISBN 978-3-8394-4147-3

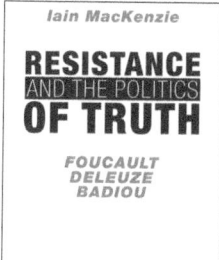

Iain MacKenzie
Resistance and the Politics of Truth
Foucault, Deleuze, Badiou

2018, 148 p., pb.
29,99 € (DE), 978-3-8376-3907-0
E-Book: 26,99 € (DE), ISBN 978-3-8394-3907-4
EPUB: 26,99 € (DE), ISBN 978-3-7328-3907-0

**All print, e-book and open access versions of the titles in our list
are available in our online shop www.transcript-verlag.de/en!**